Everything You Always Wanted to Know About

# HEATING WITH WOOD

"Wood warms you twice, once when you cut it and again when you burn it."
—Henry David Thoreau

Everything You Always Wanted to Know About

# HEATING WITH WOOD

by Michael Harris

*Photography by Sally May Harris*

THE CITADEL PRESS • *Secaucus, New Jersey*

*To Larry, Theresa and Bettina, who kindled the flame*
*To Daisey, who makes it glow forever warm and bright*
*To Mother Earth, who nurtures us all.*

First edition
Copyright © 1980 by Michael Harris
All rights reserved
Published by Citadel Press
A division of Lyle Stuart Inc.
120 Enterprise Ave., Secaucus, N.J. 07094
In Canada: General Publishing Co. Limited
Don Mills, Ontario
Manufactured in the United States of America

Designed by Dennis J. Grastorf

Library of Congress Cataloging in Publication Data
Harris, Michael, 1950-
　Heating with wood.

　1. Dwellings—Heating and ventilation. 2. Stoves,
Wood. 3. Fireplaces. 4. Wood as fuel. I. Harris,
Sally May. II. Title.
TH7437.H37　　697'.04　　79-26127
ISBN 0-8065-0718-7
ISBN 0-8065-0686-5 pbk.

# Contents

     An Invitation   7
       Let's Get Serious About Wood   11
(1)  How the Forest Produces Fuel   14
(2)  Managing the Woodlot for a Perpetual Fuel Supply   22
(3)  The Art of Woodcutting   33
(4)  Chain Saw Safety   44
(5)  Getting It Out of the Woods   46
(6)  Seasoning Firewood With the Help of the Sun and Winds   52
(7)  Finding and Buying Firewood   58
(8)  A Cord Is a Cord . . . Or Is It?   65
(9)  How Many Cords Are Enough?   67
(10)  Planning a Home With Wood Heat in Mind   71
(11)  Installing the Wood Heater   82
(12)  How a Woodstove Works   91
(13)  Fireplaces   94
(14)  The Chimney   103
(15)  The Fire   114
(16)  Cooking on a Woodstove   124
(17)  Cookstove Recipes   133
(18)  Completing the Cycle   136
(19)  How to Select the Woodstove That's Right for You   141
(20)  How to Buy a Chain Saw   147
(21)  Wood Splitters   151

## THE BUYER'S GUIDES

(1)  The Woodstove Buyer's Guide   157
(2)  The Chainsaw Buyer's Guide   186
(3)  Tree Identification and Heat Value Guide   188

Much of the credit for the information contained in this book—especially the consumer's guides—belongs to Susan Salls and Richard Wright, publisher and editor of the *New Hampshire Times,* the statewide weekly newsmagazine.

As one example of their ongoing commitment to in-depth editorial coverage, the *Times* supported the extensive research for, and originally published, this most comprehensive Wood Stove Buyer's Guide ever produced, an ambitious attempt to help consumers make sense of the bewildering array of woodburners on the market today. The *Times* also provided the inspiration for much of the other material assembled here, and its annual "Heating With Wood" Special Edition continues to be a definitive source of up-to-date information on developments in the wood heating field.

I would also like to express my gratitude to pathpointers and friends whose perspective, diligence and assistance in one way or another helped to make this work a reality: Murray Levin, Paula Span, Michael Zizzi, Terry Lewko, Lorette Pickard, Bob LaPree, Rose Cravens and many others who shared their expertise, especially a certain farmer here on Loudon Ridge.

# An Invitation

WHAT IF SOMEONE approached you while you were sitting around the house one weekend, and offered you a pleasant family adventure in the countryside. Suppose this mystery man promised you some vigorous exercise coupled with a new, refreshing form of recreation in the sunshine and crisp air, with striking scenery all around.

If you're like most people today, you'd probably first begin to wonder how much it was going to cost you, who was behind this scheme, what the gimmick was, and what this person stood to make on the deal. Now suppose you were told that instead of being charged admission, this entire adventure was free, and moreover, by participating, you and your family could *earn* $50 to $100 or more a day. *Sure,* you might say . . . or you just *might* find yourself standing in line.

The fact is, you don't have to wait around in your backyard for this stranger to arrive—the opportunity is available to you today. The place is the National Forests and many state forests scattered around the country, and the activity is cutting wood for heating fuel. No less than the President of the United States himself is behind the plan, and the gimmick is to get you to consume less non-renewable, ever-dwindling fossil fuels. While you're saving money, the nation as a whole will benefit from a stronger economy, a better environment, and even an improved forest resource!

For a great many people, the idea of heating with wood conjures up nostalgic memories of Father or Grandfather

# [8] HEATING WITH WOOD

stoking up the old parlor stove with split oak logs in the dead of winter, or perhaps recollections of the delicious aromas of home-baked bread or apple pies wafting from the kitchen cookstove. Those were the good old days, these folks might reflect, but—they usually add—those are also days gone by. The stoves, the splitting hammers, the old woodshed and the other trappings of wood heating have since become memories in the museums of our minds. Today most of us still reach for the thermostat on the wall to ward off the cold.

This practice may have caused little hardship when fossil fuels were plentiful and relatively cheap, in those halycon days when the supply of cheap oil seemed both endless and secure. By the best estimates of government and the oil industry, however, those days *are* gone forever—ushered out in a frenzy of gas lines and soaring home heating fuel prices by that great awakening, the original energy crisis of 1973, when a few bearded men in long, flowing robes tightened the valves on the supply lines that literally run America. If there was little other benefit that resulted from that oil embargo, at the very least, we were treated to an eye-opening glimpse of the changes an

*Make it a family affair!*

oil-scarce future could deliver to our energy-dependent mode of living.

Of course, it doesn't have to be that way—if we get serious about alternative energy sources before the world's petroleum reserves dwindle even further. It is true, as nearly all Americans suspect, that the short-term supply of oil is manipulated with the profit motive in mind, but as Ralph Stow, a top executive with one of the world's largest oil companies recently noted, "There are only so many beans in the petroleum jar. It is bound to run out eventually, and pretty soon the oil that is left is going to be too valuable to burn."

Energy experts say that the petroleum binge that spawned nearly 75 years of unprecedented, even undreamed-of, industrial and social development and consumer affluence is just about over. Where do we go from here?

One of the answers is back to the nation's forestlands. While nuclear power, once viewed as a rosy energy alternative, has fallen far short of its promise, and the best efforts of government and industry have not yet resulted in the development of a simple, economical solar technology available to all, the fuel of our forefathers is, if anything, even more suited to today's than yesterday's world. The technology required for burning wood is simple and readily available. Properly managed, the fuel is a renewable resource that need *never* run out.

There is plenty of wood to be burned, and thousands upon thousands of people are rediscovering the solid practicality and distinctive pleasures of heating with wood each year. Whether they use wood as a primary or supplementary source of heat, bankers and back-to-the-landers, farmers and city folks alike are reaping the physical, economic and even spiritual benefits, associated with wood heating.

Nearly everyone can save money by the efficient use of wood as fuel, while deriving satisfaction from personally providing for one of life's basic necessities and helping to improve the national quality of life.

There are other benefits as well: Heating with wood means independence from heating fuel shortages, removing the fear of freezing during the winter. It offers the opportunity for many days of stimulating, varied physical work throughout the year, a chance for normally desk-bound people to get their backs into their living. It also presents an unusual invitation to family togetherness, for working up wood is a project that can engage every family member.

Yes, heating with wood does involve quite a bit more work than turning up the thermostat, and it can demand some adjustment in weekend work habits and lifestyle, but probably much less of each than many newcomers to the field might expect.

To do it enjoyably and safely, using wood as heating fuel requires a bit of education, bringing together the forgotten knowledge of woodburning's past and up-to-date information on the best present-day developments. That education is the purpose of this book. Altogether, *Heating With Wood* is designed as a complete, thought-provoking, how-to-do-it-and-why guide for the uninitiated and aspiring woodburner, to make the experience of warming your home with wood a safe, satisfying, profitable and enriching one.

And if you do have any further questions . . . please, don't hesitate to ask.

—MICHAEL L. HARRIS

*Loudon Ridge, New Hampshire*

# Let's Get Serious About Wood

TODAY, the American economy and the continuing functioning of our industrialized society—as well as our international balance of payments deficits, our national economic crisis and our most pressing environmental problems—are all inextricably tied to oil. From many different corners, these factors are coalescing at a critical stage, even as our oil supplies dwindle. Unlike the phantom energy crisis of 1973, however, the present crunch is here to stay—at least through our lifetimes—government planners say.

It is against this backdrop that modern wood heating is coming of age. According to a just-released U.S. Department of Energy survey, the sale of woodstoves for residential heating has increased by a factor of five in this country since 1972, and more than a million homes in the United States are now heating primarily with wood. In states like Maine, Vermont and New Hampshire, *more than half* the population currently uses wood as a primary or supplementary heating fuel. These people have already discovered what thousands more will learn in each of the next few years: Heating with wood is practical, economical, ecological, and good for the body and spirit.

According to the U.S. Environmental Protection Agency, wood heat is also "safe, available, cheap, renewable and versatile."

These characteristics are strikingly illustrated by some unique applications of wood as fuel. In October 1977, a municipal public utility in northern Vermont became the first utility in the United States to generate its electricity by burning wood. Now, The Burlington Electric Department—with the overwhelming approval of local voters—is constructing a new, wood-fired generating plant in the center of their city.

Because wood chips have already replaced oil as fuel for some of Burlington's electrical generators, electric rates have actually gone *down* in that city, and future rate reductions are planned for each of the next five years. According to the department's general manager, Robert Young, the new plant will burn nearly 1,000 tons of wood chips per week, derived from "weed trees" that are unmarketable as timber. "This harvesting has the same effect on forest land as weeding has on a garden," one state forestry official observed. "The harvesting in no way erodes forested areas. In fact, it actually promotes the growth of healthy timber."

In Michigan recently, a state government task force concluded that the state could generate its *entire electrical supply* solely by burning "dead and decaying wood fiber, logging residues, mill wastes, forest thinnings and surplus annual growth, with ample megawatts to spare." In Georgia and North Carolina and other states throughout the nation, governments, utilities and private industries are looking to the forests for new sources of fuel.

Wood fuel is also relatively clean. According to the EPA, widespread burning of wood instead of coal would eliminate 75 percent of generated airborne ash. Other research laboratories have concluded that burning wood instead of oil and coal would reduce atmospheric sulfur dioxide and carbon dioxide emissions—both of which scientists say pose grave dangers to the future health of our entire planet. Wood burning is the speeded-up process of wood decay.

Despite these advantages, wood enegry is still a sadly underused resource, and government policymakers by and large seem to be ignoring its potential. According to the U.S. Forest Service, most of the annual forest growth in the United States is not utilized; it simply dies and rots on the forest floor. This same agency tells us that nearly 50 percent of the nation's forests consist of "weed trees," unsuitable for practical use as lumber, but quite satisfactory for use as fuel wood. Contrary to popular belief, agency officials say the amount of hardwood forest in this country is actually on the rise and that woodlands now cover fully one-third of North America. Quite literally, then, the economies of Canada and the United States could run on wood energy, while the trees, properly managed, could warm our homes for endless winters to come.

Right now, though, Americans, comprising 5 percent of

the world's population, use about one-third of the world's annual production of energy and other non-renewable resources. In fact, Americans waste more energy than the total amount used by the poorer half of the world's population, who, incidentally, rely primarily on wood as an energy source. For the good of our nation as well as the rest of the world, we must cut down on our consumption of fossil fuels, and one of the easiest, most painless ways to accomplish this is to use wood for home heating. For example, a recent University of New Hampshire study estimated that by 1985, wood could replace about 45 percent of the costly heating oil now used in New England, if the incentives are there.

There are encouraging signs that officials in high places are beginning to recognize the potential of heating with wood. At this writing, Congress is considering a tax credit to encourage people to purchase woodstoves, and legislation that would bolster the supply and marketing of wood for home heating. Meanwhile, however, individual homeowners are carrying the ball alone, making this potential come to pass on their own. Let's get on with the job.

# (1) How the Forest Produces Fuel

EVERY DAY, even as we scurry around the planet in search of new heating fuel supplies, the great solar furnace high above us provides enough natural energy to warm all the inhabited places on earth. Some scientists are carrying the energy quest into ever more exotic nuclear and laser technologies, but many others are casting an eye back to The Source, pondering the possibilities of harnessing the energy emanating from the sun.

Heedless of the designs of backwoods technohippies, alternative energy engineers and Western civilization in general, nature has developed its own method for collecting and storing solar energy—and the process has been underway for at least the past billion or so years. This energy conversion is the basic life process of the planet, and it's not likely to end any time soon.

Here in North America we are especially well endowed with a vast productive solar energy farm, billions of acres in size, the major "wood bank" of the world. It's difficult to go anywhere in the U.S. without seeing some of the solar collectors; they grow in city parks, along neighborhood streets and in the rolling countryside. Nature's solar energy batteries are our trees.

Of all the earth's living things, only green plants have the ability to absorb the energy of raw sunlight, lock it up in molecules of organic matter, and store it for future use. The process is called photosynthesis, and because it provides the basic building block of stored energy on earth, plants are the foundation of all higher forms of life. This is the meaning behind that puzzling but popular bumper sticker that asks, "Have you thanked a green plant today?"

Though we are most familiar with the food energy stored in our vegetables and the heat energy recovered from trees by burning wood, the energy chain that begins with plants reaches into every aspect of our day-to-day life. The gasoline that powers our cars, the oil or coal that generates electricity, and the gas we use to cook our meals are all forms of stored solar energy. These fossil fuels began as great fern forests whose molecules, with their energy stores intact, have been compressed in the bowels of the earth for more than a million years. The entire Industrial Revolution, then, was financed by old solar energy locked in molecular vaults. Soon, however, this energy inheritance will be depleted, and as the end draws

*Wood for industry . . .*

near, attention is once again focused on that most highly developed of nature's own solar collector: the deciduous tree.

Since primitive algae first began the photosynthesis process, nature's solar fuel cells have continuously adapted, self-selected and evolved. Primitive forms declined as the more efficient and productive models prospered, each suited for life in its particular environment. Over countless generations, trees have perfected their systems for manufacturing organic chemicals with energy collected from sunlight. The fruit of this evolution, the crown of arboreal creation, covers the hills and valleys and plains all around us: the nation's hardwood forests.

As solar energy collectors, deciduous, or hardwood, trees are superbly structured for the job. Firmly rooted in the earth, the sturdy trunks filled with solar storage areas hold a lofty crown of broad, flat, sunlight-collecting leaves skyward.

*Our trees are Nature's solar batteries.*

Though trees customarily grow in direct opposition to the gravitational force, their greater impetus is to seek sunlight; the branches will lean and grow toward any available bit of solar radition. Best of all, nature's solar collectors reproduce themselves, assuring a perpetual supply with only limited cooperation from humankind.

It is an early midsummer morning, and the sun rises over the eastern hills, the treetops reverberate with activity that is invisible to the naked eye. The sunlight has not yet penetrated to the forest floor, but already the leaves are sorting out the energy-rich wavelengths of the sun's white light. Just beneath the surface of the leaves is a layer of fuel cells containing chlorophyll; it is here that the miracle of transformation from energy to matter takes place. The chlorophyll triggers the basic chemical reaction, bonding the sun's energy into organic molecules consisting of carbon dioxide and water. This reaction proceeds continuously inside every healthy green leaf in the forest during the time the sun shines upon it. Even as photosynthesis occurs, the leaf "breathes in" carbon dioxide through tiny pores and exhales life-giving oxygen to the rest of the world.

Half of this newly collected energy is needed to fuel the life processes of the tree, just as the tree's complex life-support systems are essential for the continuation of the photosynthetic process itself. A steady supply of water is needed for photosynthesis, and this water supply, along with essential nutrients from the soil, is absorbed by roots buried deep in the earth and pumped upward through pipelike vessels to the delicate veins, or vascular bundles, that both nourish and support the lofty leaf cells.

In return, the leaves send their synthesized energy back through another set of vessels to the roots and trunk of the tree. Some of this energy is stored as starches for future nourishment; most of it, however, becomes cellular tissue—primarily cellulose wood fiber and the gluelike lignin that holds the fibers of the tree together.

The needle-shaped leaves of the pines, spruce, hemlock and other evergreen trees perform this energy conversion process in exactly the same way. More primitive evolutionary forms than their broadleaf counterparts, however, the evergreens are not as efficient in their collection of solar rays.

As winter snows become spring freshets in the northern woodlands, swollen tree buds burst forth in a colorful display of male and female flowers. For a few weeks between the bare branches of winter and the full green crowns of summertime, these flowers drape the tree's stark silhouette in a gauzy cloud as millions of pollen grains from the stamens of one tree ride the breezes in quest of the female flower of a nearby neighbor. Inside the female blossom is a vulva-like organ called a pistil. Once fertilized, the ovule inside the pistil develops slowly in its sheltered, womb-like surroundings, until that autumn day when the seed drops to the ground as a fully ripened acorn, chestnut or twin-winged maple seed.

Another winter must pass before the seed pokes its first tiny, tentative shoot through the mantle of fallen leaves that sheltered it from the view of marauding chipmunks, birds, and squirrels. The odds are overwhelmingly against a single seed reaching even this state; even fewer seedlings will complete the long life and tremendous growth that transforms the spindly sprout to its fullest potential as a massive, mature tree.

As a tree grows, fueled by the life-giving process underway

*Maple buds*

in its leaves, the one-cell-thick cambium layer adds new wood as annual rings. Thus the bark must expand to continue its protective role around the stem's increasing diameter. In most species, the outer bark consists of a layer of dead cells, which crumbles into a corky substance, or—in birches—peels off in layers when split by expanding wood inside, revealing another layer of bark beneath. This new bark layer then undertakes the primary protection role, until it too, toughened and dry, yields to yet another cycle of shedding, growth and renewal.

The tough skin of a healthy tree protects the tender tissues within from insects, fungi and other infestations. Barring injury or illness, a young tree growing in full sunlight, rooted in moist, fertile soil, will rapidly increase in size while synthesizing its full quota of solar energy. This pace of growth will

*Maple bark*

continue until the tree reaches the maximum mature size for its species. The period of intense growth may range from as little as 50 years for birch and poplar to as long as 100 years for stately oaks. After maturity the growth rate will slow, though centuries may pass before the tree succumbs to the ravages of insects, disease, windstorm, fire or timber cutting.

To consume energy for its life functions, a tree extracts it from molecular storehouses by a process called respiration. Oxygen is combined with the organic molecules, resulting in a release of carbon dioxide, water, the chemical energy used in the tree's cells, and heat. Energy can also be released from the wood in another way, yielding carbon dioxide, water and heat, by oxidizing or burning it. These processes are quite similar. Though burning produces considerably higher temperatures at a much more rapid rate and involves the breakdown of the cellulose and lignin, both are methods of releasing all of the energy stored in the tree's cells by photosynthesis.

In a typical hardwood forest, the total annual production of wood and bark in tree stems and branches is estimated at two to three tons of dry organic matter per acre. At 7,250 BTUs (British Thermal Units) of usable heat per pound of dry wood, this translates into an annual net energy production of 30 million to 40 million BTUs per acre.

When harvesting nature's solar collectors to release the stored energy in a woodstove—that is, cutting firewood—it's helpful to realize that a sustained yield wood supply can more easily be gained by working with the forest than against it. A well-managed woodlot can produce one cord of firewood per acre per year, on the average, but the keys to insuring a perpetual wood supply are selective cutting and conservation.

One of the surest ways to strip the woods of its future lumber and fuel supply is to seek out the straightest, tallest young hardwoods for feeding the wood stove. Sure, they'll buck up into convenient-sized two-foot lengths and they'll split easily, but trees under 12 inches in diameter that have managed to pierce the forest canopy and reach their leafy crowns to the sky are only at the threshold of their greatest period of growth. In another decade or two (the blink of an eye by forest management standards), that spindly red oak or sugar maple may become a 26-inch sawlog worth 100 times its value as firewood. Even if you are not managing your woodlot for a future timber harvest, it's best to cut healthy, dominant specimens *after* they've reached their stage of maximum growth. At this point, the trunk will measure about 22 to 26 inches in diameter.

In the average timber cutting operation, more than 50 percent of each tree cut—stumps, limbs and branches—is usually left on the ground to rot. This may be practical for large-scale operators, but there is no reason why this wood, or even sound, fallen branches from the forest floor, cannot be converted into fuel for the parlor or kitchen stove. Though we live in a land of forest plenty, it's interesting to muse, while collecting limbwood, that the most serious fuel shortage in the world today does not involve oil, but wood, which more than half the earth's population uses as its sole source of heat.

While the superior heating qualities of maple, ash, beech and oak are well known, it's not necessary to abandon the birch, poplar and even pine growing in your woodlot. You will probably want to rely on the heavier woods for midwinter warmth, but the lighter woods, though they have less heat value per cord, certainly have their place in the well-managed cordwood pile. They may not hold a fire overnight, but these woods are well suited to producing quick cooking fires or to taking the chill out of the air on a late fall or early spring day.

"Too many people who cut their own firewood don't know enough about forestry to use good harvesting techniques," Paul Bofinger, president of the Society for the Protection of New Hampshire Forests, recently observed. "Not only do they cut the wrong trees, wasting good saw timber and leaving a less valuable stand, they tend to cut trees too early." If you are having trouble developing a plan for the wisest use of nature's own solar energy system growing on your land, your county forester will provide on-the-scene guidance, free of charge.

# (2) Managing the Woodlot for a Perpetual Fuel Supply

ONE OF THE MOST important—and often neglected—parts of the homestead is the woodlot. It's often taken for granted, but good tree-producing land can be valuable indeed. In fact, in terms of its production value the properly managed woodlot may be the subsistence homestead's—and the aspiring back-to-the-landers'—most important asset. But the key here is *active* management, based on selective timber cutting and consideration of the future's needs. The nation's forests *can* provide plenty of fuel for us and future generations, but to do this it is imperative that people using the woods for fuel become at least part-time tree farmers.

In many ways, the forest is like a vegetable garden. Nature plants the seeds, but after that the laws of competition take over—and no matter what anyone tells you, that can lead to impoverishment for all. Without the gardener to weed, thin and manage the woodlot for its intended purposes, a bountiful harvest is by no means certain.

What are the uses of the small-scale woodlot? Firewood is the forest product on most people's minds today, but to go into the woods flailing with an ax at the first trees in sight is to squander the multi-use potential of most forest stands. Proper selection of trees for firewood, however, will invariably result in improvement of the woodlot, growth for future firewood as well as other forest products. Just a few acres can provide a virtually unlimited supply of firewood.

Other goals associated with small woodlot management include cultivation of trees for eventual timber or pulpwood harvest, the development of a sugar maple orchard for the production of maple products, providing a suitable environment for wildlife, maintaining or enhancing forest aesthetics and even growing Christmas trees. None of these goals is exclusive of the others. You *can* have your firewood and your forest too. What the homestead forester needs first, though, is a plan.

Take a walk through your woodlot, a leisurely walk so you can examine things in detail. What tree species are present? Which are dominant? At what stage of growth are most of the trees in the stand? Are most of the trees young and crowded together, crooked, gnarled and twisted around one another, or straight and well-spaced, with adequate room for growth? How much woodland does your property contain? With this information in mind, you can begin to formulate a plan for the management of your woodlot. Given the tree species available, what do you want out of it?

Naturally, you'll want firewood to heat your home, and, of course, hardwood is best. This should be gathered from "cull"

*An immature but overgrown hardwood stand in need of thinning.*

or "weed" trees, leaving the "crop" trees that will prove more valuable when put to other uses.

"Don't worry about having enough trees to cut for firewood," advises New Hampshire forester Kevin Richardson. "There will be more than enough." Many of the species that are least valuable as sawlogs or pulpwood, such as beech and hophornbeam, have the best relative heat values as cordwood. It is comforting to know that in the forest, as in all of Nature, equanimity applies.

A woodlot consisting mostly of three- or four-inch saplings requires that about 200 of them be cut to produce a cord of wood, whereas a 25-year-old stand of 18- to 20-inch-thick trees may produce a cord of wood from just two trees. An old Wolf beech tree can even provide a couple of cords by itself, so you'll want to work with the biggest trees, and save the smaller ones for the future.

In dry, sandy soils the pines usually do better than other species. If you are interested in a pine stand for reasons of aesthetics or for a future timber harvest, it might be wise to initiate your cordwood cutting by weeding out all other trees.

*After selective thinning, trees have plenty of room to grow, and sunlight can penetrate to the forest floor, nurturing new growth.*

If a pine forest does not fit into your plans, you can manage for the oak family, which produces high quality timber as well as firewood suited to the coldest winter nights, and also flourishes in this type of soil. Weed species and trees which are unsuited to dry soils but will make good firewood include the aspens, the birches, pin cherry, elm, ash, sugar maple, beech, basswood, hickory, hophornbeam, sassafras and willow.

Moist loamy or clay soils generally support a wider variety of tree species, and result in more productive woodlots. Among the species that flourish in these sites are red oak, sugar maple, yellow, black and white birch, ash, basswood, black cherry, red and white pine, hemlock, white, red and Norway spruce and larch. Those that might be considered weed species include aspens, elm, pin cherry, hophornbeam, blue beech, sassafras and willows.

In a pastoral setting of forest land and fields in Canterbury, New Hampshire, Tim Meeh and Jill McCoy manage the woodlands of Meeh's family homestead for a number of uses. Their woodlot is larger than most—nearly 1,000 acres—but the principles are precisely the same, and they don't spend a penny for the entire winter's heat.

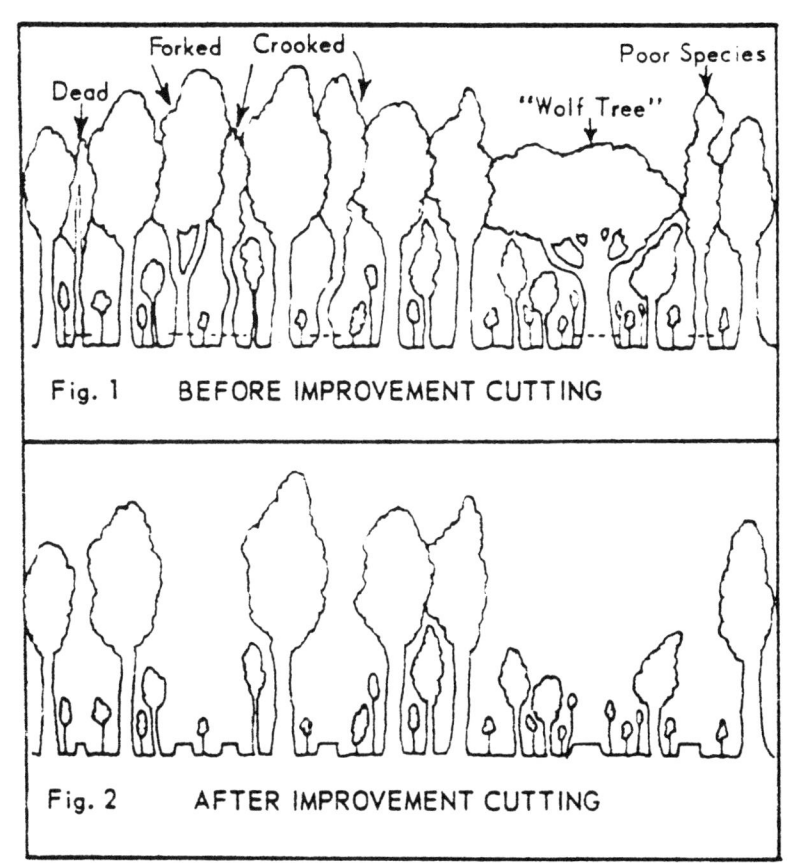

Fig. 1  BEFORE IMPROVEMENT CUTTING

Fig. 2  AFTER IMPROVEMENT CUTTING

Courtesy U. S. Forest Service

*Selective cutting in the forest stand. Here dead, diseased, crooked, and undesirable trees have been removed, leaving a well-thinned, two-story, many-aged forest stand with plenty of space for growth.*

"This is a typical cross-section of the northern forest," Tim Meeh said as he walked along the gravel path that leads into the woods. "A mixed conifer-deciduous stand. We have some young, ten- to twenty-year-old stands that you can hardly walk through yet, and some mature, eighty- to a hundred-year-old stands that are ready for timber harvest. In between those extremes are a lot of stands that need improvement cutting. That's where you have to make some decisions."

You are supposed to be able to get one cord of firewood per year from an acre of land on a sustained yield basis. This means that a productive five-acre woodlot, teamed up with an airtight stove, should warm *any* energy-efficient house virtually forever. "What the New England woodlands need most of all is to be thinned out," Tim Meeh noted, "so the high quality trees can grow. Of course, the junk wood makes good firewood."

This isn't the case in many parts of the country, however. In much of the southern and midwestern forests, for example, heavily harvested woodlots *must* be replenished, usually by planting seedlings. (Your county extension service can tell you if any of the several government reforestation programs or your state nursery offers low-cost, bulk hardwood seedlings in your area.) At any rate, grazing cattle, goats and other livestock in the woods will absolutely prevent reforestation, and should be avoided.

Most of their active woodlot management is done in winter, when Tim and Jill are finished with other seasonal farm chores. "We take advantage of the snow and ice to drag out trees from the woods," said Tim. "They slide easily and don't damage the frozen ground." With a tractor, Meeh hauls whole trees out of the woods to a landing where they are cut up into sawlogs and firewood, eliminating extra handling of the wood in the forest. "Horses will do the job well, too," he said.

"I walk into a wooded area and look around," Tim said of his management technique. "I look carefully and take my time. I look for trees with straight stems, and the most desirable species. I cut around trees like these, giving them plenty of room to grow. If I see a crooked tree, though, or one with a lot of dead branches, I mark it for cutting regardless of the species. If you find a tree that is decayed, crooked or with a fork low to the ground you know it will never make anything but firewood. You should cut it right away, because it is probably shading a good tree that is trying to grow nearby."

"Woodlot management is a personal thing," Jill added. "You have to use your own sense of things."

"Sometimes I tend to cut too lightly," Tim said. "But that's okay, because you can always go back a couple of years later and cut again if you need to."

"For people just beginning forest management I highly recommend calling the county forester," Jill concluded. "They are quite helpful, showing you examples of which trees should be cut and which should be saved, right on your own land. They know what the markets are for different wood products, and they will come out free of charge."

This service is available throughout the United States.

With your management objectives firmly in mind, heavy boots on your feet, chain saw bar to the rear, and protective devices in hand, you walk into your woodlot. Where do you begin?

Examine the group of trees before you. Desirable species,

*Each of these maple trees is hindering the growth of the others. Most of them should be harvested for firewood, leaving only the best, straightest specimen for future timber.*

like oak, maple or birch with straight stems larger than 12 inches in diameter make high quality saw timber. These should be saved, to add more growth—a 20-year old tree is just reaching its maximum growth period—or to be harvested in quantity so they will be easily salable. Your firewood supply should come from the remaining trees if you plan a future timber harvest, both from undesirable species that are weeded out and from competing trees that are thinned.

If you are planning to manage your subsistence woodlot for firewood only, the same principles apply. To maximize the growth of a single tree you may have to cut six or more trees around it—generally, all those that come in contact with its uppermost branches or "crown." As a rule of thumb, leaving about 20 feet between each hardwood trunk will provide plenty of space for future forest growth. Cut roughly an equal amount of "undesirable" species and trees that simply need to be thinned each year for your winter fuel supply. Most importantly: Plan for a forest stand that has trees in every stage of growth, to assure a smooth, continual, sustained crop of firewood.

It is quite possible to harvest a cord of firewood per acre per year from your woodlot, but to do this you must also cut dead trees and collect fallen limbs and branches. Don't forget pruning. Pruning the lower branches of trees improves the quality of the lumber they will yield, and the branches will supply kindling if nothing else. The tops of trees cut for lumber should also be used for firewood. Anything that is not soft and punky can be burned.

If you're not willing to harvest your woodlands this intensively, a half cord per acre per year is a more realistic sustained yield estimate.

Insect-infested trees should generally be salvaged as soon as they are discovered, both to prevent further loss of the wood and to halt the spread of the infestation. Rotten tree sections, low forks, deformities such as bumps, swelling and twisted stems also indicate likely candidates for the cordwood pile, but before you put the chain saw to the wood, there are the factors of aesthetics and wildlife to consider.

The woodlands are an important part of the homestead not only for their tangible yields but also for the enjoyment they provide. If a particular tree or forest stand especially appeals to you just the way it is, don't cut it. It may be worth more to you growing than in the woodbox. (Don't get carried away with this idea, however, lest you find yourself freezing while savoring the view. In most cases you will find that cutting a spindly, twisted, bent or overbearing tree will improve the area's appearance by releasing the growth of lusher greenery in the forest's understory.)

Some thought should also be given to the wildlife habitat your woodlands will provide. A variety of age classes of trees insures a continuous supply of food and shelter for forest-dwelling animals, while areas of interspersed conifers and hardwoods will provide food and protection. Brush piles at the edge of the woods will provide cover for ground-dwelling creatures, while den trees are essential for tree-climbing species. If you spot a tree with a round hole or other deep cavity in it, chances are good that it is the home of some wild creature, perhaps a nesting area. To remove such a tree for the relatively small change of firewood would be unnecessarily cruel.

While you're at it, you'll find a well-managed woodlot is a pleasure to stroll through. Don't be afraid to plant a few crocuses and daffodils among the trees. Enjoy!

As a final incentive to improve the privately-owned woodlot, the U.S. Department of Agriculture sponsors a Forestry Incentives Program that offers cost-share help to the forest landowner. The purpose of the program is to improve the calibre of the nation's woodlands. Owners of woodland tracts from 10 to 500 acres may be eligible to receive 25 to 75 percent of the cost of reforesting or improving the forest stand

*A dark, overgrown forest (left) becomes a sunny, pleasant, productive woods (right) after selective thinning.*

from the federal government. County foresters can furnish more information about the Forestry Incentives Program, in addition to providing valuable guidance to the small-scale woodlot owner.

Contrary to popular belief, the U.S. Forest Service says the amount of hardwood forest in the nation is actually on the rise, with woodcutting becoming an increasingly common practice in the nation's woodlands.

It is difficult not to feel positive about the current trends in small woodlot management. The benefit to the nation's woodlands is mirrored in the healthful exercise and flourishing spirit of self-reliance of the people involved. By using wood instead of fossil fuels for heat we are both lessening the environmental burden of air pollution and putting off, however slightly, the day when offshore oil drills will dot the New England coastline and the Western Plains will be strip mined for their underlayer of coal. America's international balance of payments is enhanced by the use of our homegrown, renewable fuel instead of oil, and the resulting insulation from energy shortages is bound to warm the heart.

*A squirrel raised her young in this tree. The old dead apple tree is almost useless as firewood, but it's priceless to the creatures that inhabit it.*

in the forest's understory.)

Some thought should also be given to the wildlife habitat your woodlands will provide. A variety of age classes of trees insures a continuous supply of food and shelter for forest-dwelling animals, while areas of interspersed conifers and hardwoods will provide food and protection. Brush piles at the edge of the woods will provide cover for ground-dwelling creatures, while den trees are essential for tree-climbing species. If you spot a tree with a round hole or other deep cavity in it, chances are good that it is the home of some wild creature, perhaps a nesting area. To remove such a tree for the relatively small change of firewood would be unnecessarily cruel.

While you're at it, you'll find a well-managed woodlot is a pleasure to stroll through. Don't be afraid to plant a few crocuses and daffodils among the trees. Enjoy!

As a final incentive to improve the privately-owned woodlot, the U.S. Department of Agriculture sponsors a Forestry Incentives Program that offers cost-share help to the forest landowner. The purpose of the program is to improve the calibre of the nation's woodlands. Owners of woodland tracts from 10 to 500 acres may be eligible to receive 25 to 75 percent of the cost of reforesting or improving the forest stand

*A dark, overgrown forest (left) becomes a sunny, pleasant, productive woods (right) after selective thinning.*

from the federal government. County foresters can furnish more information about the Forestry Incentives Program, in addition to providing valuable guidance to the small-scale woodlot owner.

Contrary to popular belief, the U.S. Forest Service says the amount of hardwood forest in the nation is actually on the rise, with woodcutting becoming an increasingly common practice in the nation's woodlands.

It is difficult not to feel positive about the current trends in small woodlot management. The benefit to the nation's woodlands is mirrored in the healthful exercise and flourishing spirit of self-reliance of the people involved. By using wood instead of fossil fuels for heat we are both lessening the environmental burden of air pollution and putting off, however slightly, the day when offshore oil drills will dot the New England coastline and the Western Plains will be strip mined for their underlayer of coal. America's international balance of payments is enhanced by the use of our homegrown, renewable fuel instead of oil, and the resulting insulation from energy shortages is bound to warm the heart.

*A squirrel raised her young in this tree. The old dead apple tree is almost useless as firewood, but it's priceless to the creatures that inhabit it.*

HEAT VALUES

*Heating Values Per Cord of Air-Dry Wood*
*Available Heat Units*
*(in million BTUs)*

### BEST

| | |
|---|---|
| Locust, Black | 26.5 |
| Hickory, Shagbark | 25.4 |
| Hophornbeam (Birch) | 24.7 |
| Oak, White | 23.9 |
| Beech | 21.8 |
| Maple, Sugar | 21.8 |
| Oak, Red | 21.7 |
| Birch, Yellow | 21.3 |
| Ash, White | 20.0 |

### MODERATELY GOOD

| | |
|---|---|
| Maple, Red | 19.1 |
| Tamarack (Pine) | 19.1 |
| Cherry, Black | 18.5 |
| Pine, Pitch | 18.5 |
| Birch, White | 18.2 |
| Maple, Silver | 17.9 |
| Pine, Norway | 17.8 |
| Elm, White | 17.7 |
| Birch, Grey | 17.5 |

### POOR

| | |
|---|---|
| Hemlock | 15.0 |
| Spruce, Red | 15.0 |
| Butternut (White Walnut) | 14.3 |
| Cherry, Pine | 14.2 |
| Aspen, Trembling | 14.1 |
| Fir, Balsam | 13.5 |
| Willow, Black | 13.5 |
| Pine, White | 13.3 |
| Basswood | 12.6 |

NOTE: A cord of sugar maple contains about 22 million BTUs (or British Thermal Units, a measure of heat energy.) This is equivalent to the energy contained in about 150 gallons of heating oil, 180,000 cubic feet of natural gas, or 3971 kilowatt hours of electricity!

*Hillside erosion: A classic example of what happens after clear cutting, in which the forest growth that holds the topsoil in place is stripped away.*

# (3) The Art of Woodcutting

WHETHER YOU'RE PLANNING to begin burning wood for the romantic touch it adds to chill winter nights or to make serious inroads on your heating bills, you're on the threshold of an education, a new experience and a lot of exercise if you intend to collect the fuel yourself. There's a hale and hearty woodchopper inside every three-piece business suit, a competent cutter and gatherer behind every kitchen apron. Grandpa and grandma did it, and so can you. All it takes is hard work, practice, the right information and a good deal of old-fashioned common sense.

In recent years, an aura of romanticism has crept into the nation's rediscovery of returning to wood for warmth, and this accounts for at least some of woodcutting's new appeal. The work is both practical and enjoyable to be sure, but it also presents another, more demanding facet. Sitting in a rocking chair warming your feet by the wood stove is the reward earned (and, some say, made more satisfying) by the task of cutting, stacking and hauling wood in from the cold.

Cutting wood is great exercise, and a person who relishes rigorous physical workouts will probably find that situps and jogging are no match for the strain and satisfaction of "putting wood by" for the winter. On the other hand, working up wood is probably the most demanding and the most dangerous task the average twentieth-century civilized person will undertake, so it's important to be both mentally and physically prepared.

Here's common sense rule number one: Don't start out for

the woods without a clear mind and a sense of your own physical limitations. You can feel great at the end of the workday, using muscles you never knew existed before, but continuing past the point of fatigue toward sheer exhaustion will certainly increase the risk of serious injury and probably sour you on the idea of future woodcutting as well. (If you have doubts that you should be working in the woods at all, it's a good idea to see your doctor first.)

Woodcutting—like any other high art—encompasses judgments, values and sensibilities that defy capture by the printed word. There's no substitute for hard-earned experience in the woods, and because of this, most professional foresters recommend that neophytes spend some time working into the woods on their own. Education, though, is the first step to doing the job and doing it right, so for the newcomer and Saturday morning hacker alike, here is a concise, practical introduction to the art of woodcutting.

The first step is the choice of tools. Romance aside, there is a strong case for using hand tools in the woods: The ring of an ax blade and the rasp of a bucksaw biting into wood add to the woodcutter's quiet appreciation of the work, while the roar and smoke of a chain saw tend to jar the senses and shatter the forest's tranquility. Today's ax and bucksaw are much sturdier and more efficient than the hand tools of a hundred years ago. However, the measure of a day's work with ax and saw might be a cord of firewood, while a chain saw, skillfully handled, could yield twice that amount of fuel. So if you don't have a lot of free time to devote to your wood harvesting project, you may have to forgo some of the aesthetics and take chain saw in hand. If you do, be sure to invest in the proper safety equipment, including ear plugs or at least wads of cotton for your ears; otherwise, you may end up grumpy from a headache rather than glowing with well-being after a day's work in the woods.

Even if you can't manage all your woodcutting by hand, you can still strike a compromise that will allow you to relish some of the woodland's serenity. Do all your chain saw work one day, and then haul the wood home or return to your cutting area later for a well-deserved, more enjoyable session of limbing, bucking and splitting with hand tools.

Whether you opt for a muscle-powered or a motor-powered saw, you'll still need a good, sturdy ax for "limbing up" or trimming the branches from the main trunk and limbs of the tree. A wooden sawbuck is handy for holding cordwood while you cut it into stove-length pieces, but a metal sawbuck poses a constant threat to your saw, should you happen to nick one of the uprights while cutting. An ax may be useful in splitting kindling-sized chunks, but anything larger can be split more quickly and easily with a splitting hammer and perhaps a couple of steel wedges. (If you have to be told to keep your woodcutting equipment out of the dirt and in topnotch condition at all times, perhaps you shouldn't be in the woods in the first place.)

After you've ventured into the woods, tools in hand, and selected a tree to cut according to sound forest management principles, the first step is deciding where to fell the tree. Determine which way the tree leans, or which side of the tree has the largest, heaviest branches, and this will tell you which way the tree is likely to fall when cut. If the tree has a pronounced tilt or a lopsided crown, you'll have little choice about the direction of its fall without resorting to much extra effort and some specialized equipment. If the tree is straight and the growth is well balanced, however, try to drop it in a spot between other trees to avoid damaging them, and as close to your cordwood pile, yarding area or hauling road as possible to eliminate unnecessary labor. Practice cutting a number of small saplings that need to be thinned before tackling a big

*Useful woodcutting tools, left to right: sawbuck holding a four-foot log for cutting; gloves and hat; splitting hammer; ax; large and small splitting wedges for tough jobs. Chain saw equipment includes: safety gloves; lubricating oil; bar and chain oil; touch-up file; earplugs; and screwdriver and wrench for adjustments. In the foreground is an old but quite serviceable two-man saw. An old-fashioned one-man bucksaw leans against the cordwood pile, flanked by its more modern equivalent, the Swedish box saw.*

[36] HEATING WITH WOOD

*Judging the fall of a tree can be tricky. Trees like those on the left can be eyeballed in relation to a weighted string, held from the fingers, that serves as a plumbline. Those on the right have interwoven branches, complicating felling with the threat of a "hangup."*

tree that you wouldn't want falling on you, should you make a mistake.

Begin your cutting by using an ax to clear away brush and low-hanging branches around the tree so you have an open area for sawing. Check for "widow makers"—dead limbs that may fall on your head when the tree vibrates from the cutting. Saw them off if they're within reach, and be sure to have your hard hat on your head if they're not. Then choose an escape route away from the stump for when the tree begins to fall— a clear path at a right angle to the direction of the fall will ensure that the tree will neither drop on you nor "kick back" across the stump and crush you with its trunk. Be sure to scramble a safe distance before stopping to witness the awesome release of energy as the huge tree crashes to the earth.

The best method of felling a tree is first to saw or chop a notch in the side of the trunk that faces the direction in which you want the tree to fall. Begin the cut close enough to the ground so that you're not wasting much wood, but high enough to be sure your blade won't strike the ground or protruding rocks. To cut the notch, make a straight cut one-third of the way through the tree, then make another cut just above it that slopes downward to meet the first cut and drops a wedge of wood out of the trunk. Then begin your backcut on the opposite side of the tree, several inches higher than the notch, and angle this cut toward the notch. While cutting, watch for an ever-so-slight widening of the slit made by the saw, indicating the tree is about to fall. When you see the cut widen into a broad smile, or hear the first cracking of wood that tells you the tree is beginning to topple, make your exit—fast.

If the falling tree stops in midair, hung up on another tree, resist the temptation to kick it, push it or pull it down. This situation is probably the most dangerous you'll encounter in the woods; the possibility that the tree could fall on you is very real. The simplest and safest method of freeing the tree is to cut sections off the trunk until the remaining wood falls of its own weight. Watch out!

*Watch out for widowmakers like these. Occasionally, they earn their name, when dead, rotted branches shake loose from a tree during cutting and land on a person's head.*

After the tree is down, the next step is cutting off the limbs close to the trunk to avoid troublesome stubs that may snag on one another in the cordwood pile. Trees are usually limbed from the bottom to top, and you can cut the limbs into firewood-sized lengths as you go. Always work your ax or saw on the opposite side of the trunk from where you're standing. Limbs standing straight out in the air can be looped off easily, but those under pressure from the trunk resting on the ground may provide some problems. These limbs may spring out with tremendous force or cause the tree to roll when they are cut; the weight of the tree may bind the saw blade if cut from the top, so they must be approached differently. First, cut one-third of the way through the limb from the top, then finish the cut by sawing through from the bottom.

There is an abundance of opportunities for binding the saw blade in the cut while felling and cutting up a tree. The safest way to cope with this problem is to drive one or two plastic or wooden wedges (not steel, which may damage the saw blade) into the cut to free the trapped tool.

Though limbs and branches are useful as firewood, most of the fuel is contained in the trunk of the tree, and this will have to be "bucked" or sliced into sections short enough to handle. Most cordwood is initially sawed into four-foot lengths. This is a convenient size for handling and transporting, and allows for later resawing into popular 12-inch, 16-inch or 24-inch sizes for burning. Of course, you can buck the tree directly into pieces that will fit your wood stove or fireplace to avoid resawing, but this will mean handling more numerous, smaller pieces in the woods.

*After judging the fall of the tree: (1) cut a felling notch one-third of the trunk's diameter into the fall side of the tree; begin the back cut until the tree falls; (2) the uncut hinge wood will direct the tree's fall; (3) buck the trunk into short lengths for ease of splitting; (4) split slabs from a large trunk section, to avoid getting your wedge stuck in the heart of an old tree.*

The best method of bucking a tree that is lying flat on the ground is to make all the cuts most of the way through the trunk from one side and then, using a peevee or a limb as a lever, roll the log over just far enough to finish the cuts from the other side. If you make the cuts all the way through from one side, you run the risk of digging the saw blade into the ground. A good saw blade or cutting chain is tough enough to cut as many as 20 cords of wood without resharpening, but if you hit the soil or nick a rock with it, you'll be spending your next half-hour in the woods with a sharpening file in hand, repairing the damage.

Once your tree is felled, limbed and bucked, you'll have to split most of the sections into chunks small enough to fit in your woodburner. Splitting is hard, heavy work, but a good splitting hammer (a sledgehammer-like tool with one side of

*A proper felling notch.*

*The back cut just before the tree begins to topple.*

the head shaped like a broadax) just light enough for you to swing it for extended periods of time without fatigue, a positive attitude, crisp air and sunshine can all conspire to make it a thoroughly enjoyable task.

Shorter lengths of wood split more easily than longer ones, so you may want to saw your cordwood into stove-length sections before splitting, though four-foot logs can also be split by placing one end against a stump and driving a steel wedge into the other end parallel to the grain. The log should split by the time the wedge disappears into the wood; if it doesn't, drive a second wedge into the opposite end of the split created by the first wedge. If all else fails, whack the side of the log with

*To avoid accidents, limb the tree on the opposite side of the trunk from where you're standing.*

the splitting hammer where the bark is beginning to come unzipped, until the wood splits in two.

Trunk sections two feet long or less are more manageable and more predictable. If they are sawed so the ends are flat and the top and bottom are roughly parallel, they will stand up easily on the ground or a chopping block (useful for keeping the splitting hammer out of the dirt) and won't topple over when hit. Pieces a foot or less in diameter can be "quartered" (split in half through the center of the trunk, and then split again to make four stove-sized pieces), but larger chunks are more easily "slabbed." This is done by splitting off the first chunk of wood from the outer edge of a trunk section parallel to the bark, and then working a spiral pattern around the outside and in toward the center. Soon you'll have a pile of slabs lying around the center, or heartwood, of the tree. You may find it easier to burn the heartwood whole rather than splitting it, since this wood, especially in older trees, often seems as hard as stone.

A smooth, powerful splitting stroke is the key to productive work. A good stroke has a telltale sound as it strikes the wood;

*Buck the trunk into four-foot lengths.*

# [42] HEATING WITH WOOD

*The right way to run a bucksaw, on a sawbuck.*

*The proper way to chop down a tree using an ax.*

it feels good and it splits the wood cleanly. It's a lot like hitting a baseball: You have to keep your eye on the target at all times, swing smoothly and follow through. Once you find a smooth rhythm and a comfortable stance, splitting wood can be as challenging and nearly as much fun as Abner Doubleday's game.

Practice this several times before swinging in earnest. Stand with your feet spread as wide as your shoulders, toes in a straight line and well back from the wood to prevent an errant stroke from dropping the splitting hammer on your foot. Hold the hammer handle at chest height, then slowly raise the hammer over your head while sliding your hands down to the end of the handle, again like gripping a baseball bat. In the same fluid motion, bring the hammer down from its position directly over your head (over-the-shoulder blows are not as powerful or as accurate) blade first into the chosen spot on the wood. Repeat this over and over again in slow motion until you begin to experience the most efficient energy flow, tossing the hammer up and over your head and then pulling it downward with all the strength you can muster, to concentrate all your efforts and the weight of the hammer head on the wood.

If this method doesn't appeal to you, you can buy or rent a motor-powered splitting machine.

Whichever methods of woodcutting you choose, you can be sure that the combination of fresh air, exercise and the satisfaction of providing your own winter's warmth will do much for your body and spirit. After the work is through, you can sit back, relax in front of the fire, and bask in the knowledge that you are heating with one of the least expensive, most intelligent fuel options available to humanity today. You can save some money while learning a new skill. It means a lot of hard work, but it may be that those pleasures richly earned are those most deeply enjoyed.

*The proper way to run a splitting hammer.*

# (4) Chain Saw Safety

THEY MAY BE the best tools for working up the winter supply of firewood quickly, New Hampshire forester Kevin Richardson warns, "but chain saws are also the most dangerous machine you can operate without a license."

Whether you're a greenhorn or a veteran at working in the woods, always observe the following precautions with a chain saw:

• Read the owner's manual once, twice, three times, until you are intimately associated with the peculiarities of your particular saw. Put off heavy-duty cutting until you are thoroughly familiar with the machine.

• Wear protective equipment, including earplugs, heavy leather gloves, sturdy boots and, yes, a hardhat. Protective pads are also available to protect your arms and legs from the bite of an errant saw.

• Be sure your saw's muffler, spark-arresting device, safety guards and controls are all in good working order before embarking on a woodcutting exploit.

• To avoid damage to yourself and your saw, always carry the saw with the bar pointing behind you.

• Never carry a running chain saw while walking in the woods.

• Before beginning a cut, be sure you're in a comfortable position with your feet firmly planted on the ground.

• Beware of cutting into dirt, stones or wire, which may

break the cutting chain and send it whipping dangerously through the air.

• Avoid damaging the bar and chain by binding them in the wood. If the saw does get stuck—as almost inevitably it will—turn off the saw before trying to free it to prevent injury.

• Never use a chain saw to slash at branches or brush. A snarled or broken chain—and personal injury—could result.

• After refilling the saw's gas tank, wipe away any spilled gasoline and always move to another location before starting the saw.

• Sharpen the cutting chain when necessary, for easier and more rapid cutting. For safety's sake, clean and maintain your chain saw frequently, in accordance with the manufacturer's instructions.

*An easy-to-build mobile crib is handy for cutting cordwood. Fill the crib with logs at the cutting site, haul it back home or to the loading landing with truck, tractor, or draft animal, then cut through all the logs at the same time with your chain saw. Using the crib supports as guides, you can adjust the length of pieces; cutting larger pieces for large wood heaters and shorter ones for cookstoves.*

# (5) Getting It Out of the Woods

Okay, you've bought yourself a chain saw, you've read the owner's manual carefully, and you're out in the woodlot ready to begin cutting the winter's fuel supply. You mark the trees for selective cutting, fell them with neat strokes of your new saw, cut the wood into four-foot lengths, and stack it in a cordwood pile between two trees. Sweat glistening in tiny rivulets down the side of your face, you stand back and proudly survey the fruits of your labor: a cord of firewood. Then, as you look off into the distance and see a plume of chimney smoke rising from your house at the far edge of the woods, the question dawns: Now what do you do with it?

That cord of wood will easily weigh more than a ton dry, and in its present green condition it would no doubt tip the scales at two tons or more. Then, too, this cord may be only one-third, one-quarter or one-fifth of your winter's fuelwood supply. "Carrying it out two sticks at a time over your shoulders," one veteran New Hampshire woodsman observed, "gets old real soon." How then do you transport the homestead wood supply from the stump to the stove?

Before skidders and other motorized wood-handling equipment were developed, nearly all woodland products—saw timber, pulp logs and cordwood—were hauled out of the woods by teams of draft horses or oxen. Most wood was transported in winter to take advantage of the easy going over hard-packed snow; large loads were heaped on oversized sleds, known locally as "scoots," pulled by the team. In summer,

[46]

fewer logs could be "twitched out" or dragged over the rough, rocky ground.

Draft animals are still a viable alternative for getting the trees out of the woods today. As a companion to selective cutting, they provide one of the least environmentally harmful methods of fuel harvesting and transport. They are also adaptable to the size of the task: While the practical limit to the amount of wood that workhorses or even oxen can drag out of the forest is determined only by the traveling conditions and the size of the team, even a single horse can provide enough power to haul a winter's average firewood supply in a matter of a few days.

There is still another live-hauling alternative for the small scale homesteader or part-time woodburner—pony power. Though they are usually thought of as little more than children's pets (and therefore are priced cheaply in relation to their work potential, especially in winter when hay becomes scarce), ponies are actually sturdy little horses, usually quite willing to do their share of work. The diminutive creatures work well on snow-packed trails in the woods, and it doesn't

*Summer-cut cordwood waiting in the woods.*

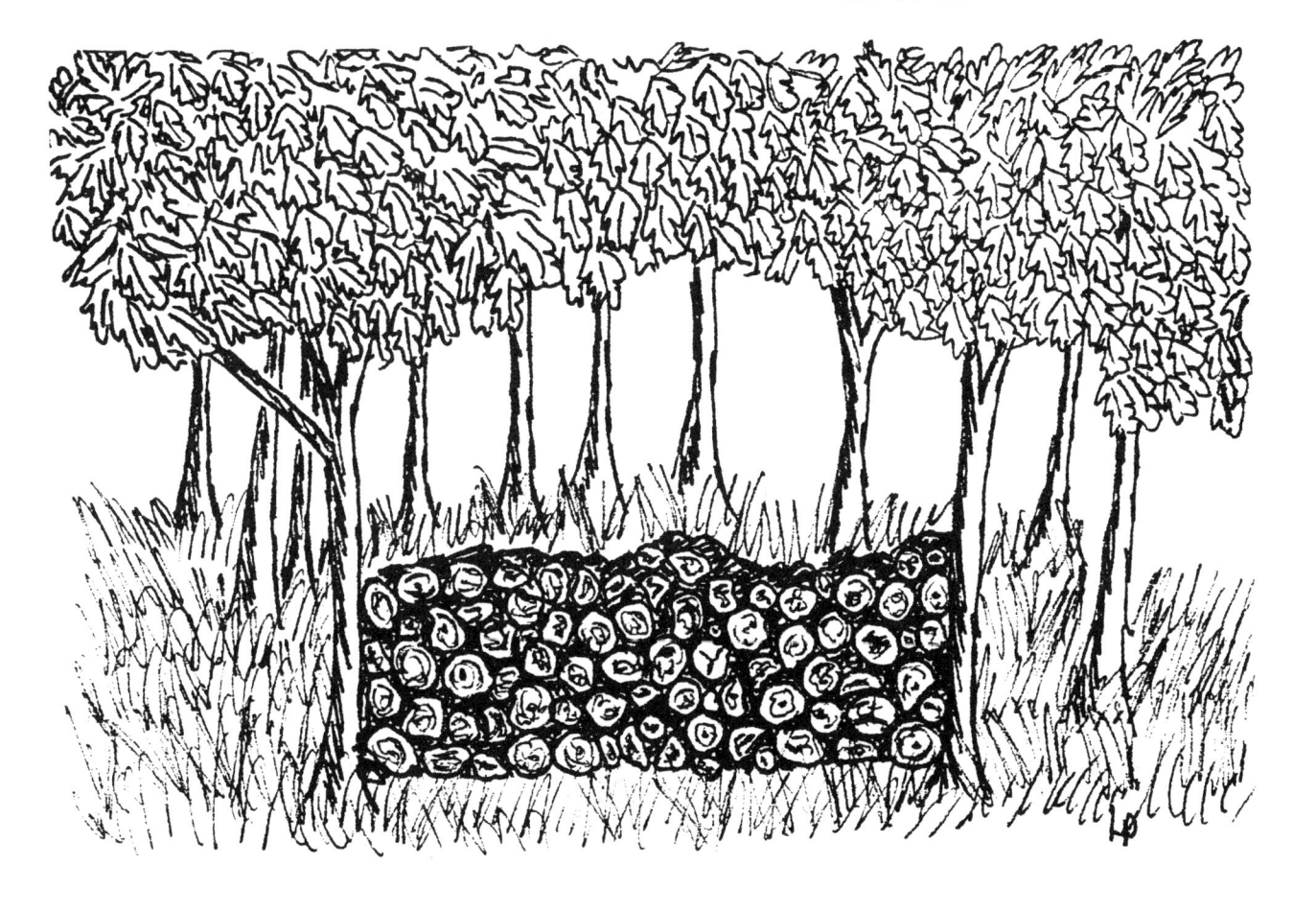

take many pony-sized logs dragged to your back door, where they can be cut up later, to amount to a cord of fuel.

But there are payments to be exacted for the keeping of draft animals whatever their size, including the initial cost of purchasing them and their hauling equipment, shelter construction and maintenance, training, grooming, feeding and general care. This is not to imply that many of these tasks aren't often as enriching and enjoyable as they are necessary but, unlike a lawnmower or tractor, a horse can't be left in the barn unfed and untended when it's not working.

Also, draft animals left idle over long periods of time often lose the work rhythm to stubbornness and an ornery temperament. "The secret to keeping a good woods horse," our wizened woodsman explains, is "working him most every day." This may mean more of a time commitment than many people have to offer.

If you're collecting wood for the winter's heating fuel but don't have a lot of spare time to devote to the project, a selection from the wide range of power equipment designed to handle the transporting work may be right for you. A tractor-drawn wagon, four-wheel-drive vehicle or the smallest skidder you can find might be right for you. All it takes is money. For the average homeowner, though, the high capital investment required to purchase this machinery to haul three or four hundred dollars' worth of fuel out of the woods each year will never make economic sense.

Should you already have a road-worn but rugged four-wheel-drive Jeep on hand, however—or have the opportunity to buy an uninspectable but still running relic at low cost—you may find that such a machine is well suited to hauling small loads of firewood along narrow trails between the trees when the ground is bare. In winter, even a family snowmobile can be made productive by hitching it up to a sled or toboggan filled with wood.

Woodcutters who already own small farm tractors may find it advantageous to outfit the machine with a small logging winch rather than a wagon for hauling firewood, particularly if they have a fairly large woodlot on rough terrain and the winch can provide an alternative to constructing expensive logging roads.

A winch mounts on the three-point hitch on the back of the tractor's power takeoff shaft. A small unit with 165 feet of $\frac{3}{8}$-inch steel cable and 6,600 pounds of pulling power may cost $1,200 or more, but it can drag out a lot of wood in a short period of time.

After the trees are felled, they are left in log lengths (or whole, if a tree is small enough and the top will be cut for kitchen wood) and the tractor is parked on the nearest woods road pointing toward the yarding area. The end of the winch cable is dragged to the butt ends of the trees, and fastened to them by means of chains or a three-pronged grapple hook, which prevents the logs from getting snagged on rocky ground. At the shift of a lever, the winch begins reeling in the cable, dragging up to four average-size trees out of the woods in a single load.

When the trees have reached the tractor, they can either be dragged to the cordwood yarding area near the house, or rewinched through the woods once again to their final destination. An average winch load is about half a cord, but with the right-sized trees a 25-or-more-horsepower tractor can haul out a cord at one time.

If you don't want to get involved with animals, and mechanized equipment, for one reason or other, is not your cup of tea, but you're still bound and determined to haul out your own wood supply, there's always old-fashioned, reliable human muscle power to fall back on. A few pieces of manual equipment can help ease the task.

Cutting cordwood into stove-length rather than four-foot sections and stacking it to dry in the woods can be quite helpful when transporting the fuel by hand. Well-seasoned wood weighs half as much as green wood, and this fact takes on critical importance when you're ferrying the fuel from the forest to the woodshed. A large construction-type wheelbarrow fitted with a sturdy rubber tire can transport a surprising amount of wood in a single load, though the size of the load and the distance it is carried depend heavily on the type of terrain and the condition of the workers involved. Large loads can be secured during their journey through the woods by outfitting the wheelbarrow with a strong rubber or nylon strap, or a light chain.

In winter, when wheelbarrow transport becomes difficult if not impossible, the family sled, toboggan or a specially made woodbox mounted on oversized runners can be put into action. There are many less pleasurable things to do than breaking a trail with snowshoes through newly fallen snow, and then early the next morning when the trail has become firm, setting off on family outing to load the makeshift lorries full of wood and send them gliding down the path to home. Again, straps of some sort are advisable to keep the load from slipping off as your cordwood caravan cavorts through the brisk winter air.

If the idea of a logging winch appeals to you but you have neither the tractor to put it on nor the inclination to buy one, there's another type of winch you can buy that needs nothing to power it but your own hands. Just as a power winch hauls

*A wheelbarrow can transport a surprising amount of cordwood in a single load, especially in the quick hands of an energetic young person.*

*Pony power: Though they're often thought of as little more than children's pets, ponies are actually sturdy little horses with amazing strength for their body size. A pony eats only one-third as much as a standard work horse, yet can pull your year's supply of cordwood home with ease.*

whole trees out of the woods with a long steel cable, a hand-driven boat winch can pull along a toboggan filled with firewood at the end of a sturdy rope. Fitted with a basic hand crank, the boat winch can be tied to a tree or attached to its own special hook in the side of your woodshed, and then reeled in to bring the cordwood load to your door. It's a physically easy task as wood gathering chores go, made even easier with the assistance of a partner to help steer the toboggan around trees and steady it over rough terrain.

Boat winches are inexpensive—$6 to $35—but it's certainly not worthwhile to invest in a cheap plastic model for such heavy-duty hauling. Hemp rope also works better than the nylon rope usually provided with the unit, since nylon cord stretches a great deal under stress and can snap with a nasty backlash when it breaks. Hemp rope, on the other hand, holds fast under tension, falls limp when it breaks and, in contrast to nylon rope, is easily mended with a trusty seaman's knot.

Whichever method you use, it's unlikely you'll forget that wood is a heavy fuel. The secret to safe, pleasurable wood transportation (and, in the last analysis, enjoyment is really what makes it worthwhile) is to haul many light loads rather than fewer heavy loads to move the same amount of wood. Whistle while you work.

*A hand winch can move a hefty load of firewood out of the forest over frozen snow.*

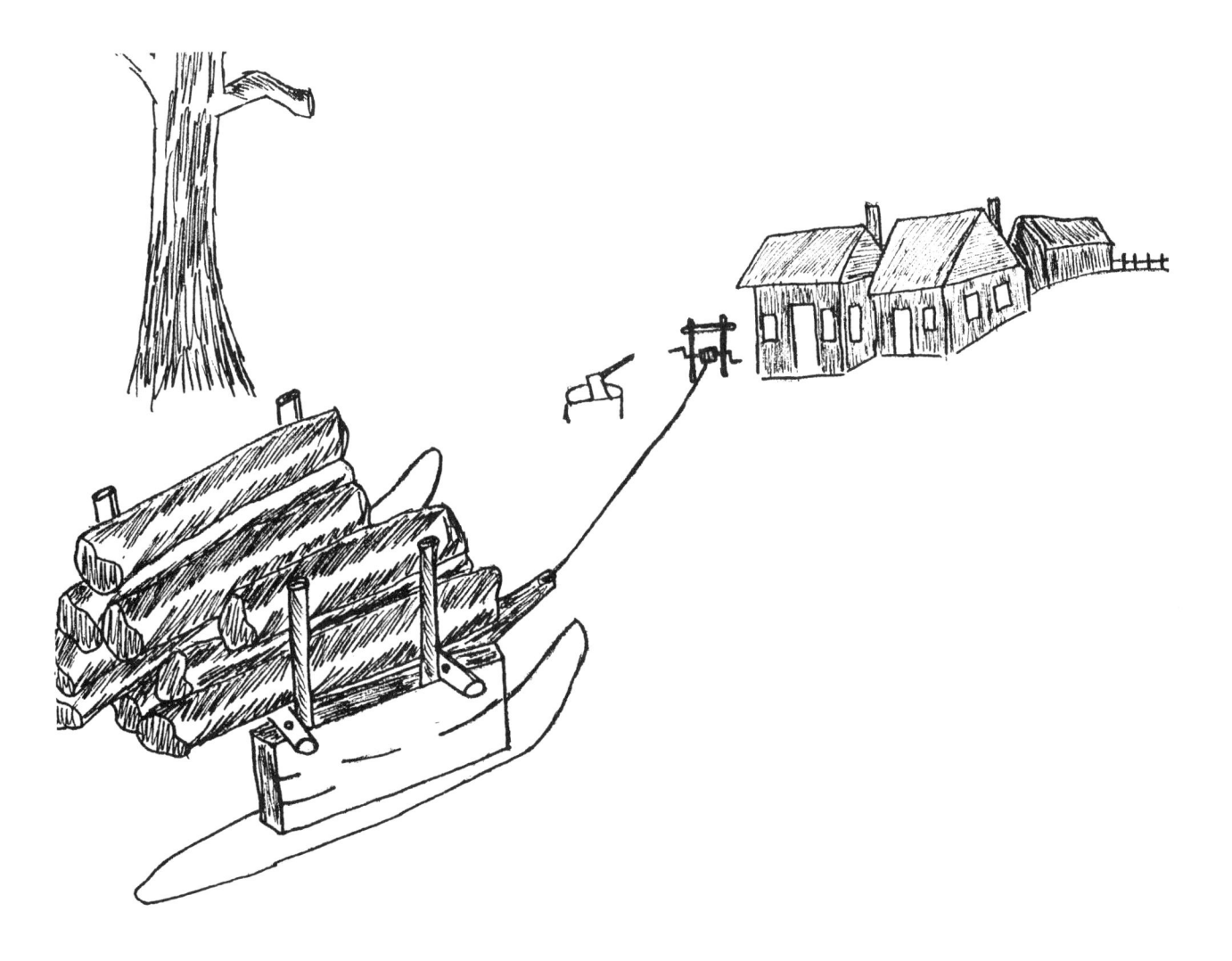

# (6) Seasoning Firewood With the Help of the Sun and Winds

JUST AS SPRING is the season for sowing and summer is the time for haying, New England farmers traditionally saved some time in the fall for cutting firewood. With the leaves just beginning to cast a golden glow over the woodlands scenery and the autumn breezes chasing away the sultry heat of summer, there's no better time to begin cutting the winter's fuel—*next* winter's fuel, that is.

Preparing the winter's wood supply—the cutting, splitting, stacking and all the hauling that takes place between the stump and the stove—is a formidable amount of work. Considering the effort involved in working up a cordwood pile—and its market value—you'll certainly want to get the most heating benefit possible out of it, and this can be accomplished only by properly seasoning the wood.

A freshly cut tree has a moisture content of about 50 percent, which is a fancy way of saying that one-half the weight of that green log you're carrying is water. This can add up to well more than half a *ton*, or from 75 to 250 gallons of water per cord.

Before the wood products in a log will burn, all this moisture must first be converted to water vapor and then driven up the chimney flue—using heat that might otherwise be put to better use warming your home. Some of this moisture also condenses in the chimney, leading to creosote problems, while lowering the firebox temperature to the point where it can no longer support secondary combustion. This further squanders the heat potential of the fuel. That is why it's best to avoid burning green wood whenever possible.

Outdoors, wood dries fastest when the moisture content of the air is lowest; that is, usually during the fall, winter and spring months between October and May. Wood cut in the autumn, therefore, will be well seasoned by the onset of the following year's winter heating season. After twelve months of seasoning, the moisture content of air-dried wood is likely to have dropped to about twenty percent—the driest it will probably get without heating the wood to drive off additional water, and quite suitable for burning—though large chunks of oak and elm may require longer drying times.

The fastest way to dry wood is to split all logs greater than six inches in diameter and cut all the sticks into stove-sized pieces immediately after felling the tree. (The shorter the piece of wood, and the more surface area exposed to the air, the faster it will dry.) If you don't have time to buck up all your cordwood in the fall, however, it's a wise idea to fell the trees, leaving them whole on the forest floor, and return to them later—but preferably *before* the winter snows have covered

*Wood drying in summer.*

## [54] HEATING WITH WOOD

*This well-stacked pile of split wood will resist winds and the passage of time without falling over, but because it has been left exposed to the winter elements, the fuel will be frozen, hard to handle, and well nigh impossible to light this heating season.*

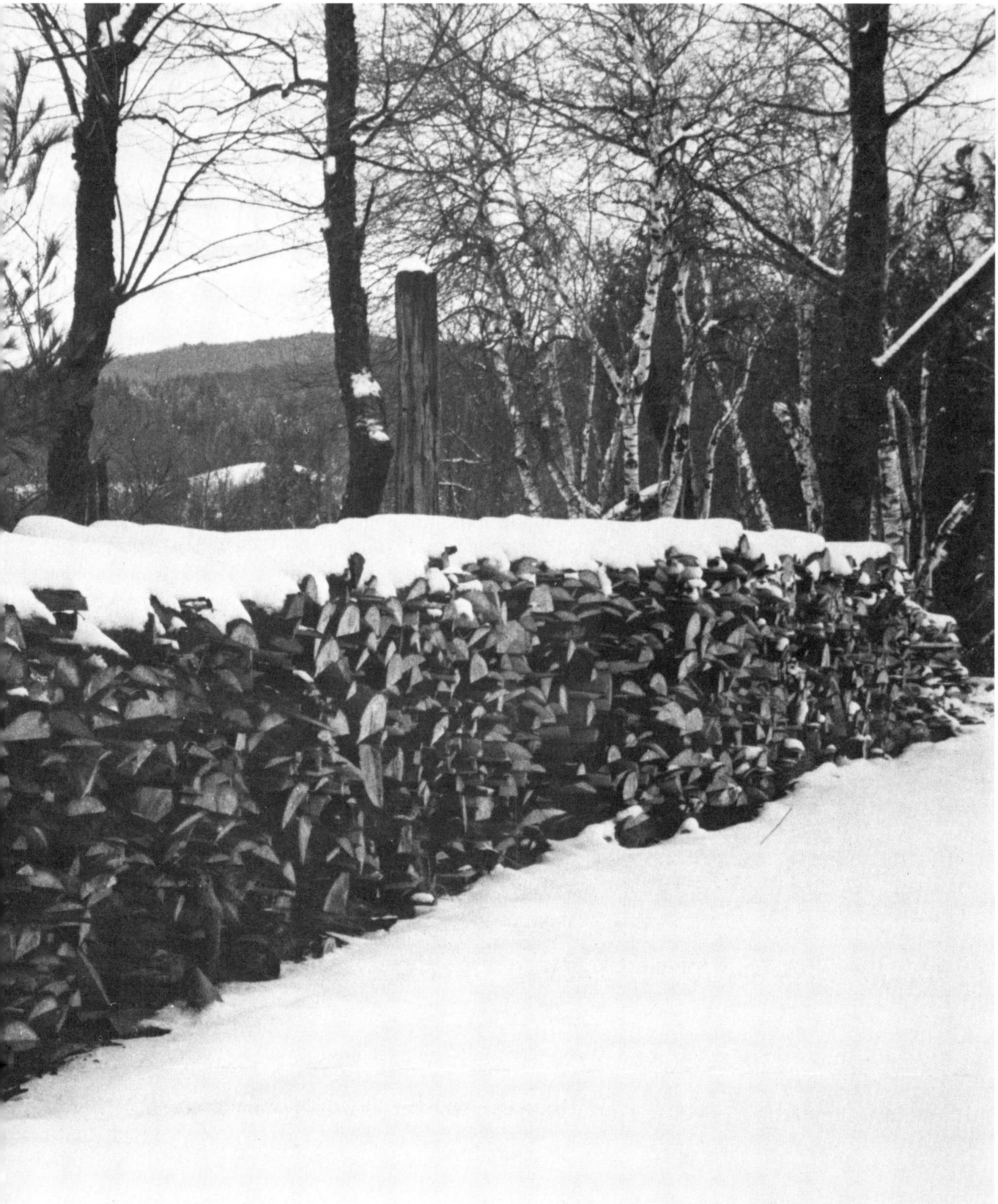

and soaked the wood, setting you back another drying season. While you are away on other matters, the leaves will be doing some of the drying work for you, pulling moisture out of the tree until they wither and turn brown. This is the time to cut the tree into cordwood.

The location of the cordwood pile will greatly affect the drying process, so select the spot wisely. You'll want the woodpile handy to your house or wood storage area, of course, but you should also select an open space free of brush and tall grass and away from large bodies of water, in order to provide for good circulation of dry air. Orient the piles so the prevailing winds will blow through the spaces between the sticks, leave two-foot to four-foot alleyways between the faces of parallel piles, and be sure to have your firewood stacked in full sunlight.

Before you begin building the cordwood pile, lay down two, four-to-eight-inch diameter unwanted logs or old timbers as runners to keep the firewood off the ground. This not only allows good air circulation, it also prevents the wood from reabsorbing moisture from the earth and eventually rotting. Stack the wood loosely, mixing split logs with whole ones and crooked and straight sticks, to allow air to move freely through the pile. Be sure to stack all split pieces with the bark facing up, to take advantage of the tree's own waterproof covering for keeping the wood supply dry.

If your woodpile will be kept outdoors in winter, it's a good idea to cover the top (but not the sides, which would hinder air circulation) with boards, metal roofing or a heavy plastic sheet. A snow-covered woodpile may be picturesque, but frozen, wet wood, is much more difficult to handle, haul and burn than dry, sheltered fuel.

To further increase the value of your hard-earned fuel, you can call on the assistance of the sun. To develop a solar-tempered wood dryer, simply stack your cordwood in a sunny location, orienting the face of the pile toward the south, and then cover the wood with clear plastic sheeting. Construct a light frame to hold the pastic about a foot off the ground on the north and south sides, allowing air to circulate freely through the woodpile, and leave spaces for air to move between the "walls" and 'ceiling' of the structure (see illustration).

The plastic will trap the sun's warmth, drawing moisture from the wood while sheltering the fuel from the rain. Under ideal conditions, the solar-dried wood should be ready to burn within six to eight months.

Well-seasoned wood is much lighter than green wood, and has radial cracks on the darkened ends of the logs. The wider and deeper the cracks, the drier the fuel. Look for these signs before you buy or burn "dry" wood.

However you season your wood, don't rush the drying time. Good firewood, like good whiskey, improves with age.

*A simple and highly effective solar cordwood dryer.*

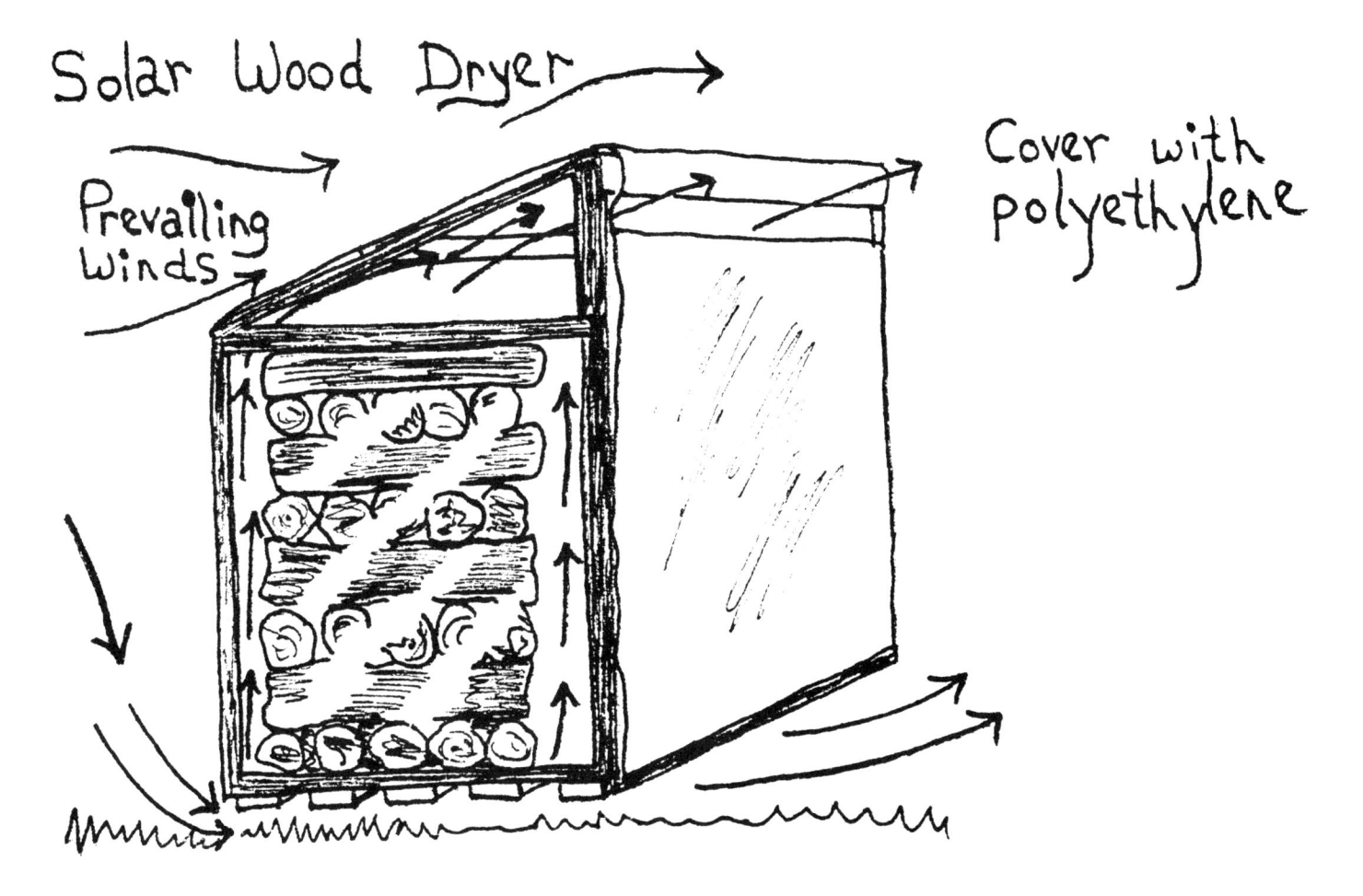

# (7) Finding and Buying Firewood

You don't have to own a woodlot or live in the country to reap the economical, ecological and spiritual benefits of wood heating. There are many other ways of gathering your own fuel supply—from finding it to buying it—in many forms. Along the way, there's lot of exercise, family recreation and fun in store for you too!

In a sense, we all own a woodlot, a vast network of productive forests called the National Forest System, supplemented by local state forest areas. Smokey the Bear's domain is full of dead and fallen trees just waiting to be collected as firewood, and thanks to directives issued from Washington, D.C., in response to the President's energy program, getting in to collect fuel to heat your home has never been easier.

The ground rules may change somewhat from time to time, but right now, every American is entitled to gather up to ten cords of firewood (for his or her own use—not for sale) free from the National Forests. You simply have to apply for a free permit from your local U.S. Forest Service Office, and observe the rules, which include cutting *only* down and and dead trees and branches. This policy is as good for your personal energy economy as it is for the nation's forests. You'll be gathering a supply of already-seasoned firewood, leaving the live trees for future woodburning folks, and the forest fire hazard will be lessened by the absence of dry, combustible material on the woodlands floor.

Many states also offer similar cut-your-own programs in

state forests, though a nominal fee is often charged for a cutting permit. For details, contact your state forestry agency.

If you live in a heavily-forested area, you doubtlessly have noticed the vast amount of limbs, tops and other tree material left behind by timber harvests. Ask permission from the operators or the landowners, and after the professional cutting is over, you may be allowed to salvage some of this firewood too.

Across the country, many people use wood as their primary or sole source of winter heat despite the fact that they personally own no more woodlands than a shade tree in their backyards. Their secret to heating with free wood successfully is to make fuel gathering part of their year-round recreational activities. You can do it too. Establish a wood-gathering rhythm, and glean your fuel from whatever resources can be found. This isn't a difficult task, but it requires a bit of imagination.

If you stop and think for a minute, I'm sure you can come up with many sources of free wood right in your own community. Remember the last time you saw a huge tree limb lying across a sidewalk, felled by a storm? Most likely, a municipal maintenance crew had to come along to take it away. City and town tree-cutting crews are also often called upon to do away with a diseased elm, or other massive tree. If you're sharp-eyed enough to be at the scene with your chain saw in hand, you can probably convince the crew members to leave the wood at the site after they've cut it to manageable size, for you to cart away. If not, check your local dump. Even in these days of high energy prices, you'll be amazed at how much potential firewood winds up there.

Whenever you're out in the country, and spot a dead tree sticking out of the greenery along the roadside or in a field, consider asking the landowner's permission to cut it down and take it home. If it's been standing there long enough to shed its bark and turn a silvery grey without being cut down, chances are you'll be allowed to have it. Whenever you drive into the woods, to hunt, fish, or gather blueberries, pick up a trunkload—or a truckload—of the dead wood you're certain to find scattered all around you. Remember: Anytime is a good time to gather firewood.

There's also lots of potential firewood you don't even have to go outside of the city to find. Mill ends, broken pallets, slabs and other lumber scraps can often be had at a local yard or sawmill, usually for free or a reasonable fee, considering the BTUs you'll be gaining. If you have the time to spend cutting and splitting the wood to size, a demolished house can also keep you in free fuel for years. My own mother grew up in a totally wood-heated urban home, and she recalls gathering the family's entire winter fuel supply from sources like these with the help of her brothers and a little red wagon.

When you do collect assorted firewood types, it often helps to segregate the various stacks in a convenient storage area, such as your basement or porch. Keep the driest woods handy for burning first, and stack the largest, heftiest pieces separately for use during the coldest winter days and nights. With a bit of practice, you'll soon develop a feeling for managing your own fuel larder.

Then of course, you can always buy your firewood. Here, too, there are a number of options.

Wood—the raw material—is in plentiful supply, but preparing it for burning is a laborious task. The price you'll pay

*Every dead tree in the woods or at the side of the road is potential firewood.*

for a cord of wood, therefore, is largely dependent on how much or how little work you're willing to do yourself. There are few real "deals" in buying firewood, but plenty of opportunities to exchange your labor for the wood dealer's efforts. You can have your wood delivered to your front door, all cut to length and split into fine pieces at a premium price, or you can buy standing trees for a nominal fee. Wood can also be purchased in many stages in between.

The cheapest method of procuring firewood of course, is to purchase standing trees, or "stumpage," from a private landowner. Wood can often be purchased for $5 to $10 per cord on the stump. Some woodlot owners are also willing to arrange an exchange program (your work for their wood) by allowing you to cut one cord of wood for yourself for every cord you cut and stack for them. If you've got plenty of time and energy but no woodlot of your own, this is an attractive alternative.

It's also fairly common for woodburners to buy cords in four-foot or even sixteen-foot lengths, then do the sawing and splitting in a convenient spot right in the backyard. Wood delivered by the truckload in this form usually costs about half that of wood already cut and split to woodstove size, but it may involve about eight hours of work per cord. If you enjoy cutting and splitting, as many people do, this arrangement will pay you handsomely for your recreation.

Many folks just aren't interested in cutting and splitting, however, and would rather trade off the ax and saw chores by coming up with a little extra cash. If your interest in wood as fuel extends only to the heating aspects, the best advice is: Shop around.

In large urban areas, the $100 cord is already all too common, and to make matters worse, city wood sometimes seems to mysteriously shrink in volume on the trip in from the countryside. If you can afford the time some weekend, drive into the country to strike a firewood deal at the source, rather than waiting for the wood to come to you. You'll probably make out better that way.

Right now, an average—and perhaps fair, considering the labor involved—price for a cord of firewood, split and cut to stove length, is about $75. (In terms of energy value, this is equal to buying heating oil at fifty cents a gallon, since a cord of good hardwood yields about 21 million BTUs, replacing about 151 gallons of oil. This equation misses one essential point, though: Wood heat is often distributed more efficiently than oil heat—except when it's released in a fireplace.)

The heat value of a cord of wood varies according to tree species, of course, but you can't simply call up a wood dealer and order a cord of oak, maple or pine. Usually, wood is cut

*Firewood comes in many forms, from many sources.*

# FINDING AND BUYING FIREWOOD

and piled as it's found in the forest, which is mixed. A variety of firewood types is handy to have on hand, but don't get caught paying premium oak prices for what is predominantly a cord of poplar. Learn to know your trees (see the Tree Identification Guide in this book), and inspect the load *before* it comes off the truck.

If you're buying wood to burn this winter, by all means be sure it is well seasoned. This means much more than asking if it is "dry." Properly seasoned wood is much lighter per volume than green wood, and is obvious because the darkened ends of the logs will have wide, deep radial cracks running through them. If you buy and burn green wood, you'll be wasting more than twenty percent of the fuel's heat value, in addition to setting yourself up for future creosote problems.

On the other hand, seasoned cordwood often demands a premium price, especially as winter approaches. The best bet is to buy your wood during the off-season, when supplies are plentiful (the end of the heating season is a terrific time), then dry it yourself for use next year.

*Tree-length logs, purchased by the truckload, are a good fuel buy.*

# (8) A Cord Is a Cord Is a Cord...Or Is It?

UNFORTUNATELY, BUYING WOOD is not a simple matter. Either you must buy your wood from a dealer whom you trust implicitly or you must become an educated consumer.

Like most other items, wood is sold by standard units of measure, and most often this is the cord. A standard cord is a volume of 128 cubic feet, usually measured as a pile eight feet long, four feet high and four feet wide. Actually, though, there is only about 75 to 100 cubic feet of wood in such a pile. The rest is airspace.

If a cord of wood consists of only large round logs, you'll be buying less firewood than if you purchase a cord consisting of many different sizes of logs, which fill in the airspaces. Examine the cord for this characteristic before you buy.

A face cord is a pile of wood eight feet long, four feet high, and as wide as the lengths of the wood cut, usually 12, 16 or 24 inches. A face cord, therefore, is always less than a full cord, and should be priced less. A 24-inch face cord is actually half a cord of wood, and a 12-inch face cord is one-fourth the full cord volume.

A "truckload" of wood is a vague description of volume, depending on how big the dealer's truck is. A standard pickup will hold only one-third to one-half a full cord, so beware if a dealer tells you he's delivering "about a cord" in a half-ton pickup truck! If you have doubts about its real volume, stack the wood and measure it before you pay for it.

Because of the non-standardized nature of the fuel, it is rea-

**Wood is more economical.**

sonable to expect small variations in the size of woodpiles. However, a wood dealer should correct substantial shortages. It is illegal to deliver wood in quantities less than those advertised and paid for. Always ask for a receipt showing the quantity of wood delivered as well as the price paid. If you think you have been cheated, call your state bureau of weights and measures.

A final note: A full cord of softwood will weigh about 2,000 pounds while a full cord of hardwood will tip the scales at about 3,500 pounds. All types of wood have equal heat value per pound, so clearly, a cord of hardwood is a much better fuel buy.

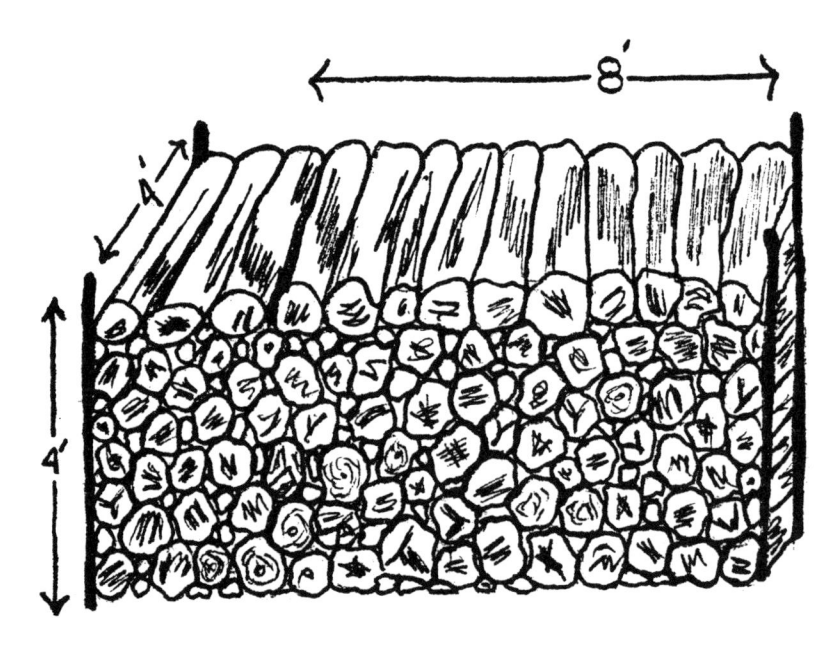

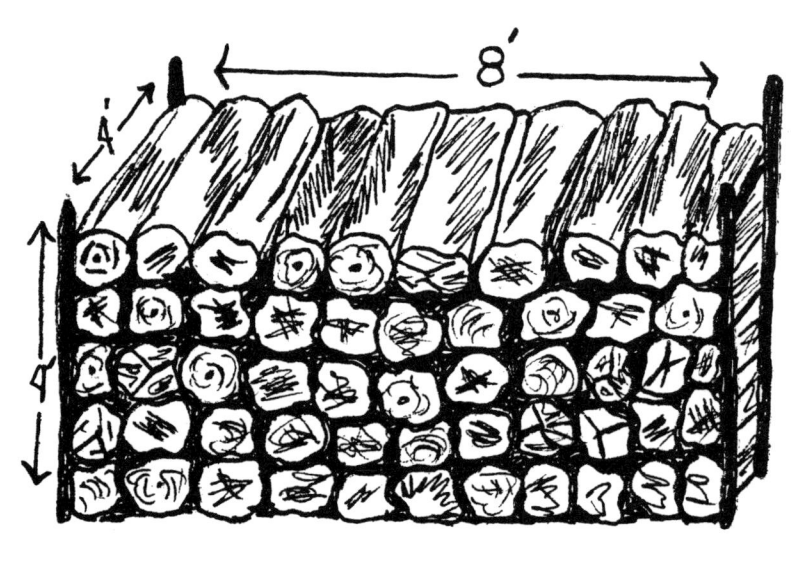

*A cord consisting of various-sized sticks contains more wood than a cord made up of large round logs.*

# (9) How Many Cords Are Enough?

AT THE FIRST HINT of northern springtime, when dark chocolate chunks of earth begin to find their way through melting snow, budding trees, warming zephyrs and even upstart crocus blossoms all conspire to draw the curtain on another heating season. If the Ides of March finds you with a cord or two of dry wood still in the woodshed, you're a fortunate one. In addition to that hard-earned sense of security, you have a head start on next fall's wood supply. But if you're like most of us, the tail end of the heating season finds you scouting out half-hidden sticks of limb wood, scurrying after those chunks that didn't seem worth the trouble of hauling last December, and—travesty of travesties—even burning a little bit of next year's still-green, autumn-cut wood.

"Next year, I'll be sure to put in enough cordwood," uncounted thousands of woodburners have been known to say at just this time each year. One problem with that resolution, however—even if your motivation doesn't wane during halycon summer days—is how do you determine how many cords are enough?

The question is even more baffling for neophyte wood users. A cord of wood seems like a formidable pile to the uninitiated, and under sun-drenched skies it is often hard to believe that a wood heater could ever consume more than two or three of them. Yet as every veteran of at least one wood-warming season knows, the icy heart of winter can whet your woodstove's appetite and convert your Bunyanesque woodpile into

[67]

a couple barrelsful of ashes.

Arriving at a precise calculation of the amount of cordwood you will need to get yourself comfortably through the winter has always been a task frought with complications. For one thing, there is the nature of cordwood itself. Unfortunately, wood is a non-standardized bulk fuel which does not easily lend itself to a precise reckoning. Unless you have a truck scale in your backyard to measure your wood in cunits, you'll probably be estimating your supply in cords—a theoretical volume equal to 128 cubic feet of wood which does not exist in reality, due to such factors as irregularly shaped logs, twists in the wood, and bark and spaces between the sticks of a 4'×4'×8' pile.

Then there are a number of other variables to consider, ranging from the type and dryness of the wood to the size, configuration and insulation of the house involved. If all of this makes the job of producing a reliable estimate of your cordwood needs seem hopeless, don't despair, for Brad Smithers has a solution.

Employed as a researcher for the Society for the Protection of New Hampshire Forests, Smithers first began working on a computer program to determine the amount of wood needed

*Happiness is enough wood to keep you warm for the winter.*

to heat a given amount of living space while an undergraduate student at the University of New Hampshire. Smithers, a veteran woodburner himself, has recently developed a formula for estimating cordwood needs based on the cost of heating the same or a similar house in previous years with electricity or fossil fuels.

"The system has proven to be pretty accurate at estimating cordwood needs," said Smithers at the Forest Society's library in Concord, New Hampshire. "Still, it never hurts to put in a little bit of extra wood to be prepared for an especially severe winter. Any wood left over will just be drier for use next year."

If you haven't had any prior experience with conventional fuels in your home, you can ask your local electric company to come out and give you an estimate of the cost of heating your house electrically. Utilities usually can develop precise cost-per-square-foot equations based on size, insulation, etc., and provide this service without cost or obligation. If you decide not to buy electric heat, you can still use this information to calculate your cordwood needs.

Using the Smithers method to estimate the number of cords it will take to heat your house with wood, all you need to know is what type of fuel you currently use or have used in the past, and how much of it you used in an average winter. By determining the amount of energy (BTUs) required to heat the house (from your last fuel bill records), it is possible to calculate the number of cords required to supply an equal amount of energy. You should also know the efficiency rating of your wood heater. This information can be obtained from your dealer, or estimated from Table 3 below.

The equation is simple:

$$\text{Cords} = \frac{B \times Eb}{W \times Ew}$$

When: B = units of present fuel consumed per year (your records).
Eb = efficiency of heating (Table 1)
W = units of present fuel contained in an "average cord" (Table 2)
Ew = efficiency of the woodburner (your information or Table 3)

An example: An increasingly impoverished and newly enlightened friend used 1,350 gallons of #2 fuel oil to heat his home during a recent severe winter. He has a standard oil furnace, but is now planning to switch to a high-efficiency woodstove. How many cords will he need to heat his house with wood?

$$\text{Cords} = \frac{1{,}350 \text{ gals.} \times .65 = 877.5}{150 \text{ gals.} \times .60 = 90} = \frac{9.75}{\text{cords}}$$

To determine the total cords you'll need for the next heating season, use the tables below and plug the corresponding numbers into the formula.

TABLE 1
EFFICIENCY OF HEATING (Eb)
Standard Oil Furnace .................................0.65
Standard lp gas/natural gas furnace ..................0.75
Electric heat ........................................1.00

TABLE 2
FUEL UNITS PER AVERAGE CORD DRY HARDWOOD (W)
#2 fuel oil .................................150 gals.
lp gas .....................................230 gals.
Natural gas ............................21,000 cu.ft.
Electricity .....................6,158 kilowatt-hours

TABLE 3
WOODBURNER EFFICIENCIES (Ew)

| Open Fireplace | .10 | Furnaces and high- | |
|---|---|---|---|
| Airtight stoves | .50 | efficiency stoves | .60 |
| Improved fireplace | .25 | Box stove | .30 |

## TERMINOLOGY

- "Improved fireplaces" are those which use some type of accessory such as built-in ductwork, a hollow grate or closing glass doors.
- "Box stoves" are non-airtight woodstoves.
- "Airtight stoves" include most new stoves with good tight seams and airtight draft controls. Air to the firebox can actually be closed off completely.
- "Furnaces and high-efficiency stoves" include wood-fired furnaces and boilers and woodstoves of double drum design or with a smoke chamber. (The high proportion of surface area makes these units quite efficient.) Stoves with more than five or six feet of stovepipe exposed in the house can also be placed in this category.

There are certain variables in each situation which this simple equation cannot take into account. Therefore, in weighing the result, you may want to consider things such as house design and its effect upon heat circulation, type and dryness of the cordwood, passive solar heat collection and, last but not least, your definition of winter warmth.

# (10) Planning a Home With Wood Heat in Mind

IF YOU ARE NOW in the process of planning, building or buying or retrofitting your dream home for back to the land, or even suburban, alternative living and hope to heat it primarily or entirely with wood, you will be interested to know that certain designs and architectural features can drastically affect the amount of space your woodstove will warm. Manufacturers who claim their wood burners will heat "two to five rooms" are not just skirting the issue. The fact is that the size, content and configuration of the rooms play a key role in determining the capacity of a wood heater.

Even such figures as "5,000 to 8,000 cubic feet" are estimates at best, since there is no way of knowing the number or type of rooms in which those cubic feet are contained. The basic rule to remember is that one large room heats more easily than three smaller rooms of the same total size. In the case of circulating heaters, this is primarily because the air flow must work itself through doorways and around corners to reach the more remote areas to be warmed. In the case of radiant heaters, walls are even more significant, since these stoves work by heating objects, not air, so the walls themselves must be warmed before they can radiate heat through to the other side —and typical stud walls, enclosing trapped air pockets, are better insulators than heat conductors.

Since objects inside a room act as heat sinks, a room full of furniture will heat more slowly than an empty room because each of these items must be warmed by absorbing radiant heat before the temperature of the air inside the room will rise. Once heated, however, this full room will stay warm longer than an empty room, as the heated objects project their warmth back into the room. More on the dynamics of this principle later, but briefly, a cozy room *will* help keep you warm.

Generally speaking, if there are two or more parallel walls between your wood heater and a part of your living space, you will not be able to heat that outlying area with a single woodstove. Rather than investing in a second woodstove installation to warm this isolated area you may want to simply install a small electric heater, especially if the outlying room will be used only occasionally during the winter months. But this problem can also be avoided entirely by "open design" architecture.

## CONFIGURATION AND SIZE

When planning, purchasing or remodeling a home with wood heat in mind, first consider the shape of the building. A house that is square or nearly square encloses the greatest amount of living space using the least possible amount of exterior wall, and that minimizes heat loss. This configuration also holds the potential for easily-developed open living space. In your planning, leave as much unobstructed space as is practical for your lifestyle. Complicated engineering calculations are not nearly as useful as personal observations in wood-heating friends' homes to establish that isolated, projecting els tend to be cold and that airy, open central-activity areas and adjacent kitchen-dining-living room areas that don't hinder air circulation are ideally suited for heating with a single stove. A large number of separate rooms connected by hallways are not, unless the heat is coming into them from below.

It's a good idea to place your bedrooms in their traditional location on a second floor if possible. Not only does this make optimum use of the roof area covering the living space (thus economizing on building materials), but you'll sleep warmer since heat tends to migrate upstairs during the latter part of the day. If you can stand the lack of privacy they afford (I cannot) open bedroom lofts boost a home's heat circulation potential, too. One kind of contemporary open design architecture, however—the cathedral ceiling—is *not* well suited to efficient heating, whatever the heat source. This is because heat tends to stack up in the overhead reaches, resulting in temperatures in excess of 100 degrees F. while floor temperatures may drop to 55. If you must have the drama of a soaring ceiling, you should have a fan up there to recirculate the heat toward the building's occupants.

When planning a dream house, the near-universal mistake of most owner-designers is to build too big. Space often seems much smaller on paper than in three-dimensional life, so make comparisons in friends' homes, and take a hard look at your lifestyle's demands and spatial requirements. Build as small as possible, striving for maximum utilization of each square foot by eliminating space-wasting hallways wherever possible, for example. The extra planning and occasional compromise will mean a much more easily heated home, and will pay dividends in building and heating savings for years to come.

If you plan to include guest rooms, a workshop or other space that doesn't need to be heated all the time, place these rooms at the cooler extremities of the structure, and fit them with tight, insulating doors that can be closed to ease the heating load when those rooms are not in use. Where often-used space is divided by partitions you may want to consider moving them. Wider-than-average doorways will also result in better heat distribution. A five-foot opening between a kitchen and dining area, for example, will produce twice the heated air circulation of a traditional-size passageway, while adding a pleasing aesthetic touch. Low-wattage electric fans can also be installed above doorways or temporarily placed in doorways to assist natural convection currents, although a variety of safety and conservationist principles can make this neither a practical nor a desirable solution.

In general, though, the longer the unhindered horizontal circulation path you can provide for the warmth emanating from your stove, the better your heat distribution will be. Don't try to accomplish long-distance heating with extensive horizontal runs of stovepipe. Such installations are difficult to clean and unsafe. In the best of circumstances, which is an unimpeded heat flow circulating horizontally and vertically throughout the living space, an efficient, properly-sized woodstove can heat an entire one- or even two-story house.

## STOVE LOCATION

In accomplishing this goal, the positioning of the stove is very important. Safety considerations come first, of course—the woodstove should be at least three feet away from any combustible surface—but the heater should also be placed toward the center of the home, away from corners, to enhance the circulation of air around it. In more spacious or sprawling houses, two stoves may be necessary: Sometimes a mainstay living room stove and a standby kitchen woodburner for use during the coldest winter days will solve the heating problem presented by a complex architectural design. *Home Energy Digest* technical director Larry Gay also suggests that radiant heaters be located near masonry or other heat absorbing surfaces and as far away from windows as possible, since a stove located near a window will transmit a great deal of its heat directly to the great outdoors.

Since heated air rises with the certainty and predictability of gravity, woodburners are more useful on the first floor than on the second floor of a two-story house. The same principle applies to a basement and first floor, except that there is so much mass in a cool masonry basement and the surrounding earth will draw so much heat away from the structure that little warmth will reach the first floor without the aid of heating ducts. Wood-burning furnaces are generally located in the basement, depending on ducts to deliver warm air upstairs,

PLANNING A HOME    [75]

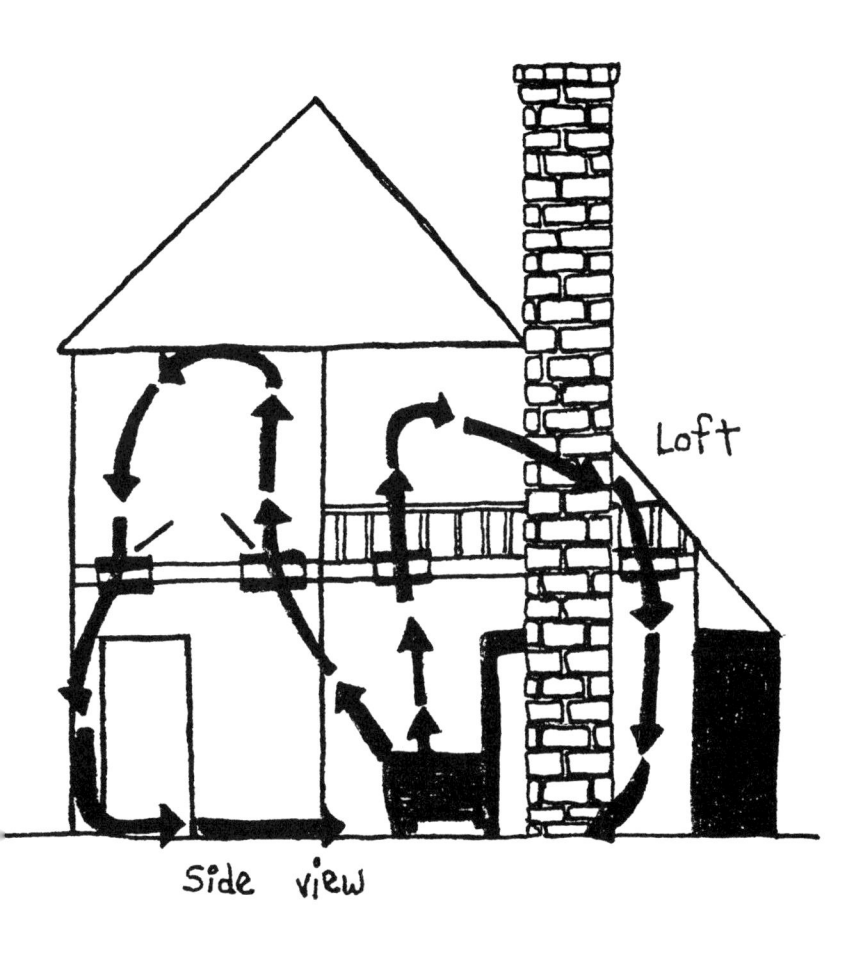

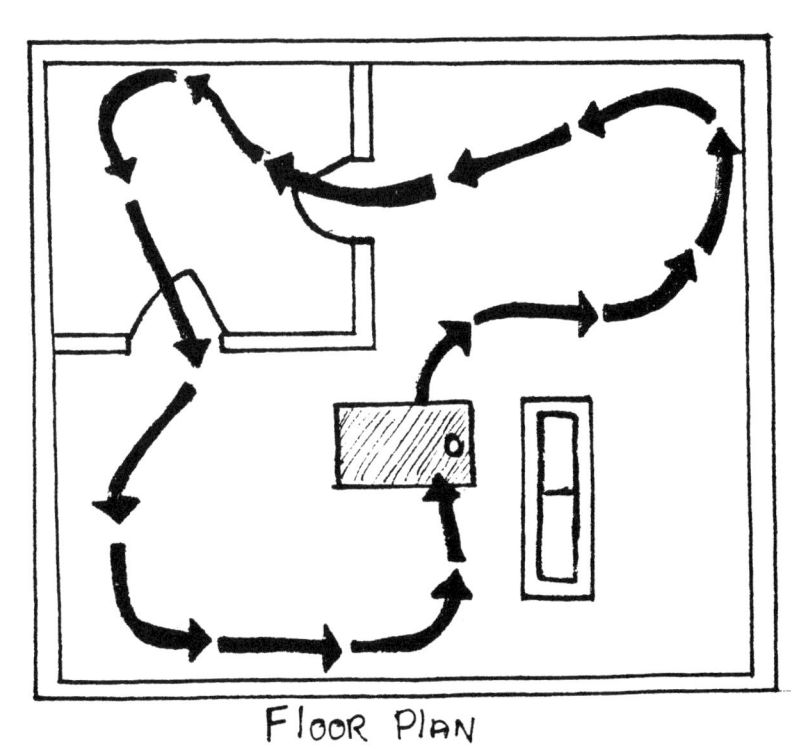

*Air-circulation patterns from a well-placed wood stove in a house planned with wood heat in mind.*

# HEATING WITH WOOD

because of their space demands and the convenience of wood handling this affords. But unless your basement is part of your central living space, you'll gain more heat benefit by locating a woodstove on the first floor.

When considering matters of convenience, by the way, one of the wood heater's primary considerations has to be the handling of the fuel. Be sure to plan your house to include *plenty* of storage space in a *nearby* woodshed or in the basement to prevent your winter's wood supply from turning into a monolithic mass of frozen fuel in the winter that slowly melts into a soggy mass that would put a fire out come spring.

Advance planning for an efficient wood transportation route from woodpile to stove is also important, not only to avoid leaving a trail of bark bits, wood chips and sawdust across the kitchen floor or having to trek across the living room carpet

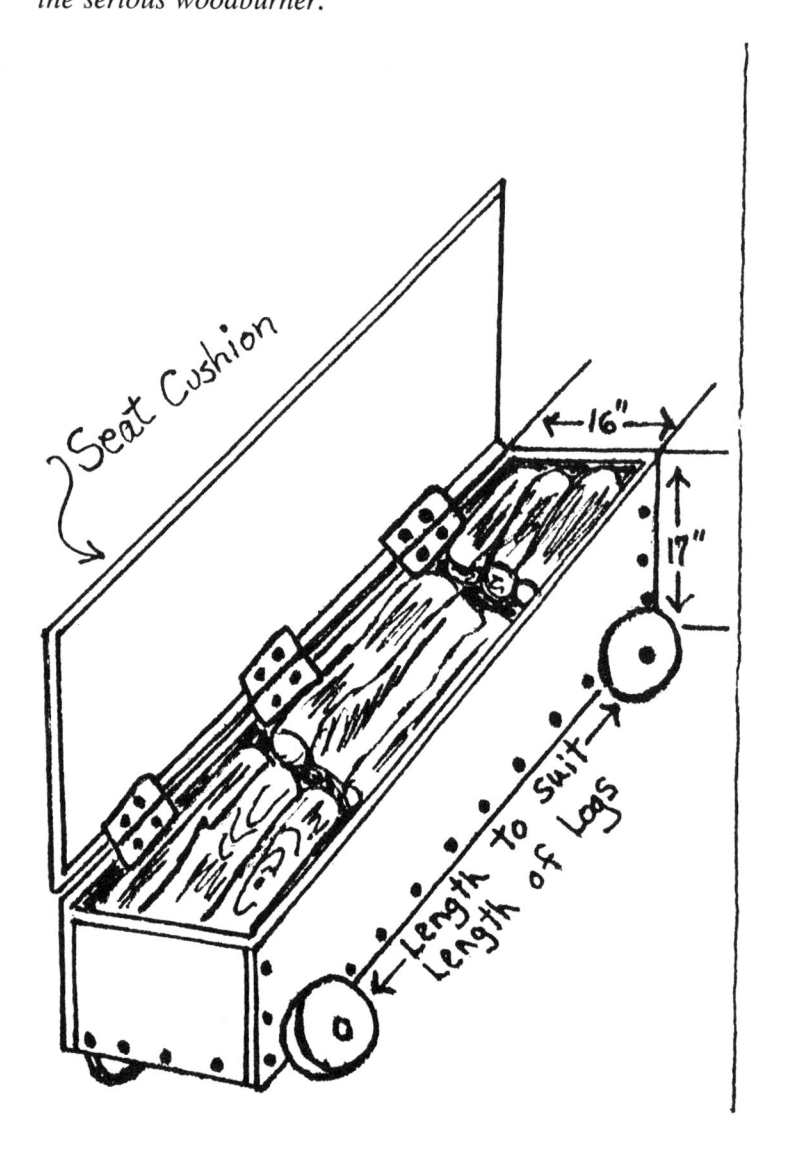

*Woodboxes on wheels and a short, direct path from stove to wood-storage area make life pleasant for the serious woodburner.*

with snow-covered boots, but to minimize the distance you'll have to carry the several tons of wood you'll burn each winter. A short back door or basement door to woodbox route is ideal. While we're on the subject, I'll confess that a friend once persuaded me—with great difficulty—to build my woodboxes on wheels. Today, when one is empty I simply wheel it to the back door (something like Mohammed going to the Mountain) and on the way back, think of this friend with gratitude for each load of firewood he saved me from carrying through the house.

To make the best use of a stove's heat, the vertical air flow to the upper floor should be regulated in some way. Heated air from a woodstove on the first floor will readily flow through an open stairwell to the second floor. This may be all right at bedtime, but it will also draw large amounts of heat into the unused bedrooms during the day at the expense of the occupants downstairs. This is the reason for the doorways found at the bottom of stairways in many old farmhouses built when wood heat was an inescapable fact of life.

Floor registers that open and close, placed between the first and second floor rooms, are effective in helping to direct the flow of warm air where it is wanted. The registers should be placed over the stove on the first floor, and they should open into the second floor rooms that are to be heated. A 6" ×6" register is usually quite sufficient to warm a well-insulated single room, but a second, return, register located near an outside wall will greatly assist heat flow. Take care when locating this return register, however, since the cooler air likely to be circulating through it will be uncomfortable for someone sitting directly under it.

If you're building a snug, open design energy efficient house, though, you may want to put off installing registers until you find out by actual living experience whether or not you'll need them. The openings between the floors do diminish the sense of privacy somewhat. With a good, powerful radiant heater, they may not be necessary.

## THERMAL MASS

Incorporating thermal mass into the house will increase your stove's ability to warm the living space, even after the fire has gone out. Furniture and any other material that will absorb heat will help some, but the best thermal mass is dense material, and lots of it. There's plenty of useful thermal mass in a masonry chimney, but as a practical matter, chimneys should never be placed outside of a house, no matter how pretty they may look to the neighbors. Such chimneys will simply pour their stored-up heat into the out-of-doors.

Most people have noticed that heavy stone, earthen or concrete structures seem cool during warm weather and warm during cool weather. One cause of this is the capacity of these materials to absorb and store heat as they increase in temperature and to release the heat again when the temperature of the surrounding air drops below the temperature of the heat-absorbing material. The heat storage quality of various materials commonly used in house construction that provide significant thermal mass can be determined by the following equation:

Heat storage capacity = Specific heat of material X Density of material

(Data is derived from the following table)

COMPARISON OF THE HEAT STORAGE CAPACITIES OF VARIOUS LOW COST MATERIALS ON AN EQUAL VOLUME BASIS

| Material | Specific Heat (BTUs/lb/°F) | Density (lb/cu ft) | Heat capacity of one cu ft material (BTUs/cut ft/°F) |
|---|---|---|---|
| Water | 1.00 | 62.5 | 62.5 |
| Iron | 0.112 | 489.0 | 55.0 |
| Concrete | 0.27 | 140.0 | 38.0 |
| Brick | 0.20 | 140.0 | 28.0 |
| Rocks (Basalt) | 0.20 | 180.0 | 36.0 |
| Marble | 0.21 | 180.0 | 38.0 |

Example: The heat storage capacity of water =
1.00 × 62.5 = 6.25 BTUs per cubic foot per degree F.

Therefore, a tank filled with one cubic foot of water will absorb—and later release—62.5 BTUs of heat for every degree change in temperature. Thus, absorption of 5 × 62.5 or 312.5 BTUs will raise the temperature of the water by 5°F. If the house is left unheated and the air temperature drops five degrees lower than the temperature of the heated water, the entire 312.5 BTUs of heat will be released to the surrounding internal environment. How much temperature stabilization this heat release will produce in any given house depends upon the location, size, insulation and tightness of the building involved. This factor can be precisely calculated by developing standard Heat Loss equations found in any basic heating engineer's manual, or heat loss can be minimized by careful use of the best design, construction and insulation methods within the reach of your building budget.

Experience has shown that the temperatures of thermal masses located adjacent or very close to the heat source can be expected to rise as much as 15 to 20 degrees over ambient air temperatures in normal household operating environments. Maximum heat storage that will be accomplished by more distant heat sinks, however, will be limited to temperature rises of 5 to 10 degrees.

What this all means in practical terms is that the masonry or other storage material absorbs excess heat from the woodstove when the fire is burning mightily, and gives off the heat, tempering the cooling air, after the fire goes out. It may take several tons of masonry material to do it, but this heat storage principle might be enough to keep your plants from freezing if you are late coming home one winter night. Incorporating an *interior* brick or fieldstone wall (on a solid foundation), a slate or concrete floor, a 50-gallon or larger aquarium) filled with life forms that can stand severe temperature fluctuations and outfitted with an emergency heater to keep the water from freezing) or, if you can afford them, marble pilasters, are some ways to add thermal mass as interesting architectural touches.

As any veteran of even one woodburning season can tell you, an ever-burning woodstove can quickly rob the air of moisture and make your home as dry as Death Valley on the Fourth of July, so make some provision to replenish the air's moisture level. This can be as simple as a tea kettle on top of the stove, or as elaborate as the bed of bricks and sand one California homeowner douses with water each day. The bed is underneath the woodstove, and the heat turns the water soaking the bricks into airborne moisture quite nicely.

## BUILD TIGHT AND SMART

Whether new home or old, conservation is the key to economical, efficient wood heating. Good insulation is nearly as important with wood heat as it is with other, more expensive, types of fuels, if only to allow your woodburner to warm a larger area. Start with tight construction and include ample insulating material; refurbish with as much energy consciousness as you can muster. Standard construction specifications now call for 4" fiberglass batts in the walls (R19) and 6" batts in the ceiling (R22). Consider this your minimum standard, and think long and hard about using six inches of insulation in the walls and up to twelve inches overhead if your budget can stand it. Wood will only get more expensive and remain just as much work to cut, haul, split and burn. And so far, the winters aren't getting any warmer.

For the woodburner, sealing the house against drafts is even more important than thermal insulation, since wind infiltration will severely affect air circulation patterns. If air leaks into your house through cracks in the walls and spaces around doors and windows, you'll find yourself freezing on the windward side of the stove while hot air roasts the folks on the leeward side. So use plenty of weather stripping and caulking compound to seal the house up tight.

Also consider using double-glazed windows, which cut in half the extraordinary amount of heat lost through glass. Even double-glazing provides only one-sixth the thermal resistance of a standard insulated wall though, so you may also want to use heavy curtains or other insulating material such as rigid styrofoam board to cover the windows at night. Finally, you may want to add or include a foyer or other "air lock" room for your winter entrance, to prevent the fifty cubic feet of cold air that pass through a doorway with every opening of a door from entering directly into the toasty warm living quarters.

Many architects now recommend venting an air supply directly to each woodburning unit, especially in very well-insulated houses. The process of combustion going on inside your stove consumes very large quantities of oxygen, which the people inside a tight house can come to miss if it is used up by a roaring fire. In fact, a severe depletion of oxygen could possibly kill you without warning.

The best procedure for insuring an adequate oxygen supply for you and your stove is to run a one-inch pipe—plastic is cheapest and quite suitable—from a hole drilled directly beneath the heating unit to the outdoors. If you have a tightly-enclosed basement, run the pipe between the floor joists to a hole drilled through the outside sill, and cover it with fine mesh wire screen to prevent summertime insect infestations. If you have an uninsulated basement or crawl space, simply run the pipe through the floor and the insulation directly beneath it, so the pipe projects into the open air.

Then you can rest easy. Not only will this pipe provide a sufficient oxygen supply and the air change that is desirable in tightly sealed houses, it will control the intake of air so that the cold incoming currents are warmed immediately upon entering, rather than drifting in haphazardly through leaks in the walls, creating uncomfortable drafts.

It's been a long-accepted bromide that homes heated exclusively with wood need constant tending during the winter months, to prevent plumbing pipes from freezing up and plants from dying off after the fires go out. (Concern not about the plants but the pipes and the house's possible resale value is the reason that your banker may require you to install a "back-up" electric heating system if you apply for a construction loan.) The newest generation of efficient, air-tight heaters, coupled with generous amounts of thermal mass, can give woodburning homeowners a 12- to 24-hour leeway for absence before anything drastic happens, but if you think you'd sometimes like to be away from home during the winter months for longer periods than that, more extensive preparations are in order.

Most household components can withstand freezing temperatures without harm (you may want to consider the fate of shampoo bottles, cans of latex paint and wallboard joint compound, the contents of your wine rack, etc.) except for plumbing systems and house plants. A small solar-heated greenhouse with adequate heat storage capacity is the best haven for the plants, although moving them to a plant-sitter's house is an acceptable alternative. Since it's as difficult to solar-heat a plumbing system as it is to move it to a friend's house, though, the best bet is to plan or adapt the pipe configuration so it can be drained easily.

What this means is simply identifying the lowest point of your plumbing system, and installing a threaded union there so the pipes can be drained before you leave, and easily reconnected for use when you get home. Another good idea is to centralize the location of all water pipes within your home, so you can warm them with a small electric heater rather than disconnecting and draining them if you only plan to be away for a day or two.

Believe me, such advance planning pays off. Not only were my mate and I able to leave our wood-heated home for a month-long trip to Mexico last February without problems, but when we are home, keeping the house at 68 degrees isn't much of a chore. Using these design methods, we built ourselves a 2,000-square-foot house covering an additional 1,000 square feet of basement (well, we didn't pay enough attention to the "build compact" advice) and it's all heated by a single woodstove. As I write this, it's terrifically windy and well below freezing outside, but here in the farther reaches of the second floor, I'm in shirtsleeves, cozy and warm.

# (11) Installing the Wood Heater

PROPER PLACEMENT of the woodburner in your home is important not only to your comfort and its heating ability, but especially to your safety. Safety is not an exciting subject, but it deserves your complete attention, because your home and the lives of your family may depend on it. Please read the following pages carefully, and when installing your own woodstove, err on the side of caution.

Old-fashioned common sense is the first requisite for woodburning safety. Unfortunately, most of us are most familiar with heating plants that are controlled by thermostats on the wall, so instinct has to be supplemented by information. Trial and error is no way to learn about woodburning safety, *so be sure you're right the first time.*

For a start, consult your local housing codes regarding woodstove installation and chimney safety requirements. If there are no codes governing solid fuel heaters in your community, follow these directives for woodstove installations developed by The National Fire Protection Association. The clearance and installation guidelines presented here represent many years of experience and provide reasonable, but not extreme, margins of safety. Use them.

A woodstove must be a certain distance away from combustible materials to prevent fires caused by radiated heat. The National Fire Protection Association recommends that the sides of a woodstove should be at least thirty-six inches away from any combustible materials. A foot- or two-foot clearance will not do. Even though very high temperatures are needed to ignite most combustible materials, over a period of time high temperatures can gradually change the chemical composition of wood so that it may eventually begin to smolder and burn at temperatures as low as 225 degrees Fahrenheit. This temperature is easily reached by an unprotected wall exposed to a stove without adequate clearance, so strict adherence to these standards are advised.

A stove may be positioned closer than thirty-six inches to a wall, but only if that wall is protected by a noncombustible material spaced at least one inch away from the wall, allowing air to circulate behind the material and carry heat away. Placing a noncombustible material *directly* on the wall offers little protection, since heat will be conducted through that material to the wall behind it, creating an even greater hazard.

Asbestos millboard (available from most lumberyards and woodstove dealers) spaced one inch away from a wall will allow an eighteen-inch stove-wall clearance. If 28-gauge sheet metal is installed in the same manner, the necessary clearance is only twelve inches. Remember, though, that these clearances are for *all* combustible materials. Avoid such dangerous but common practices such as stacking wood or paper next to a stove or moving a sofa "temporarily" closer. It may be difficult to install a woodstove where it is convenient and aesthetically pleasing as well as safe, but don't compromise on the essentials. Above all, a woodstove should be safe.

Safe floor clearances are substantially less than those for walls because the heat radiated from the bottom of a stove is generally less than from either the sides or the top of the heater. Among other factors, the bed of sand or ashes at the bottom of the stove insulates against the downward flow of heat. However, unprotected stoves have been known to generate such intense heat that they burn a hole through the floor, dropping the heater into the room below, so don't light a fire without a stove board or other protective material underneath your woodstove.

Unless the stove is on a masonry floor, when stove legs are eighteen inches or longer, it should be placed on a 24-gauge sheet of metal to reflect the heat. When legs are six inches to eighteen inches long, an approved stove board should be used, or asbestos millboard should be placed under a 24-gauge metal sheet. If legs are less than six inches in length, the stove should be placed on a layer of four inch masonry bricks or blocks covered by a 24-gauge sheet of metal.

Falling embers and sparks present an additional safety problem that is often ignored. The best way to avoid this problem is to extend the floor protection eighteen inches from the

*Specifications for safe stove installations.*

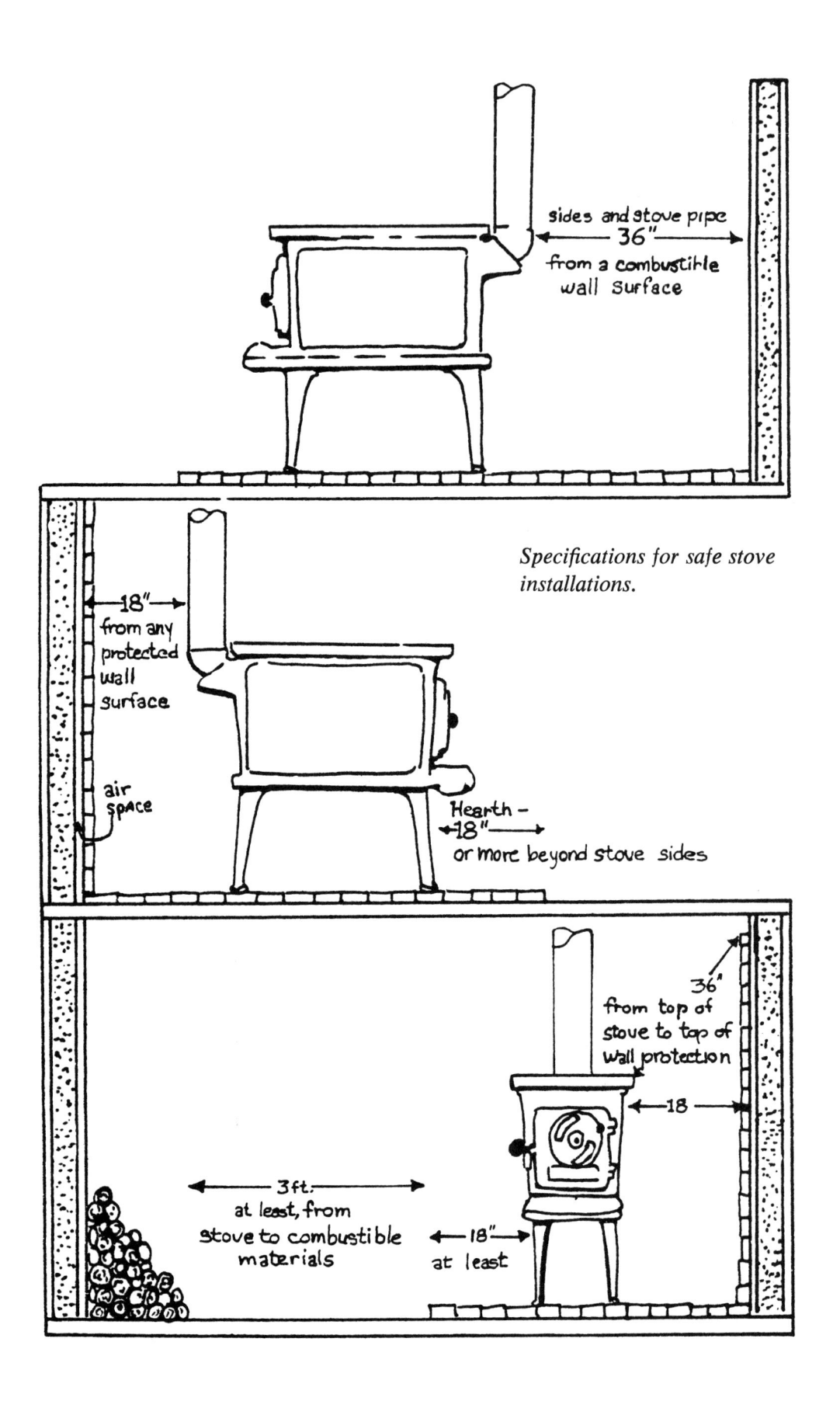

front of the stove, and six inches around the sides and back. This affords a reasonable amount of protection, but you should still take care when loading and tending the stove. Make sure that ashes and hot coals fall only on the protected area.

When the stove is installed and the surroundings are well protected, it's time to turn your attention to the stovepipe. A stovepipe must be installed so that there is a good draft to carry the hot gases away quickly and safely. This can be a tricky proposition, unless you follow these guidelines:

Keep your stovepipe as short as possible.

Keep turns and bends to a minimum.

The horizontal portion of the stovepipe should be no more than 75 percent of the vertical portion.

The stovepipe should enter the chimney well above the stove outlet to insure good draft. Horizontal sections of the stovepipe should rise at least ¼" per foot.

No portion of the stovepipe should be closer than eighteen inches to any combustible surface. If proper distance can't be maintained, then a protective shield like those described above for protecting walls from the sides of the woodstove should be erected.

Stovepipe comes in different diameters and thicknesses. Be sure to get the size stovepipe that matches the pipe connector on your heater. Pipe gauge 24 or thicker (lower gauge numbers indicate thicker metal) is a good investment because it will give better protection than thinner pipe and last longer. Stovepipe sections should be fitted together tightly. Consider permanently joining them with two or three sheet metal screws at each joint, to prevent the stovepipe from ever shaking apart and spewing smoke and sparks into the living area. This makes the pipe harder to clean, however. Inspect the stovepipe regularly, and plan to replace it when it wears thin—perhaps every two to three years if it receives constant use.

The greatest frequency of fires from heating with wood occurs at a danger point that many woodburners tend to overlook—where flue pipes pass through combustible walls or ceilings. If at all possible, avoid going through walls and ceilings, and never run stovepipe through concealed places like closets or inaccessible attics. If this is necessary, though, use an insulated ventilation thimble that is three times the diameter of the flue pipe to enter the wall (see illustration). Be sure to use asbestos-lined pipe in any concealed area. Exposed wall studs used to secure the thimble should be covered with a protective layer of sheet metal.

Improper entry to wall-covered chimneys has also resulted in many home fires. If you must pass through a wall to reach your chimney, cut back wallboard, lathing, studs and all other combustible surfaces to three times the diameter of the stovepipe. For extra protection, cover the frame of the opening with a standard pipe collar or homemade sheet metal to reflect the heat. Use a standard connector thimble to link the stovepipe to the chimney, and cement the thimble in place (see illustration).

## Wall Thimble

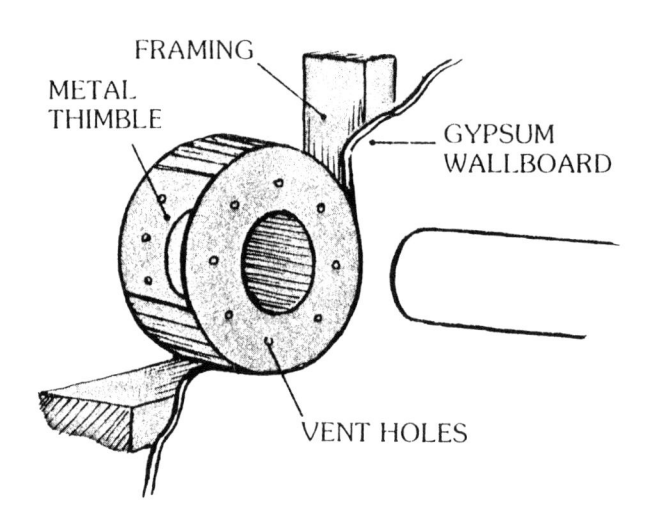

## Chimney Connection

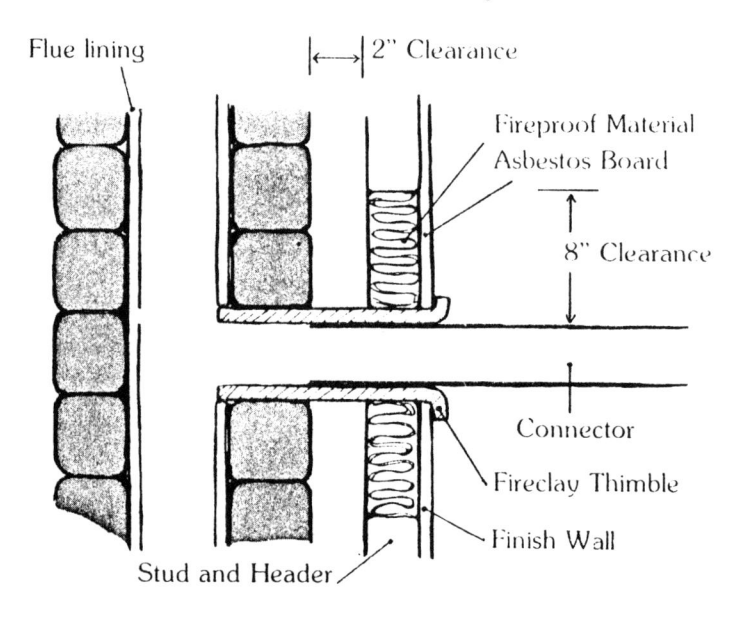

Besides a woodstove, then, you'll need the following materials to make a safe, reliable installation: sturdy lengths of stovepipe and at least one elbow, which will fit the flue opening of your heater; an in-line flue damper for all but the most efficient airtight stoves; a metal collar; an insulated ventilation thimble if you must pass through a wall to reach the chimney; a fireclay thimble and cement to connect the stovepipe to the chimney, protective stoveboard and other insulating materials if the proper unprotected clearances cannot be achieved; and of course, the chimney itself.

Stovepipes and chimneys, though they are often mistakenly used interchangeably, are two completely different things. When a stovepipe is used for a chimney, dangerous conditions may occur. Creosote may build up rapidly, and the wind and the elements may soon corrode the pipe. Consequently, if a chimney fire should occur, there will be little to contain it.

In contrast, chimneys keep the smoke and the gases hotter than does a single-thickness stovepipe, preventing rapid creosote build-up. They are also made of corrosion-resistant materials, so they do not need frequent replacement, and most modern chimneys are able to contain a fire. A stovepipe should be used only to connect a stove to a chimney; never in place of one.

It's sad but true that if your house is not set up for woodburning with an existing, serviceable chimney, it may cost you as much to buy and install a chimney as it did to purchase the stove itself. This isn't too bad, however, considering that the $400 to $700 you might pay for a complete installation would no longer cover even a single season's electric or oil heating bill in most homes nowadays.

Whatever the cost, don't skimp on the chimney. Besides enraging your insurance agent, you'll also endanger your family and home by sticking a stovepipe through a wall to conduct smoke to the outside. Do it right, and you'll sleep peacefully at night. This means constructing a new masonry chimney, or correctly installing a prefabricated, corrosion-proof, insulated metal chimney bearing the Underwriter's Laboratories (UL) certification.

Of course, many houses constructed during the last forty or fifty years were unwittingly outfitted for today's wood heat needs, by virtue of the heat-wasting but fire-resistant fireplace. The old stone hearth with a solid brick and tile flue may not be much of a woodburner on its own, but in many cases it makes a terrific site for locating a woodstove.

After checking the chimney for soundness with a smudge fire, patch any cracks, then seal off the opening to the living space with a sheet of metal, asbestos board, bricks or other nonflammable material. Set the stove on the hearth or on a stove board at least twenty-four inches away from the face of the fireplace, connect the stovepipe to the stove, and determine the proper location for a hole in the sheet metal or asbestos board to accomodate the pipe. Cut the hole, then by means of a collar, thimble and cement, fit the stovepipe to it.

There are several other devices available to make your woodburning experience safer, a few to make it easier and more enjoyable, and a great deal of gadgets aimed at capturing your dollar.

At least one smoke detector should be installed in every home, wood heated or not. (Stay away from ionizing ones; they emit radiation, no matter what anyone says. A good, large volume fire extinguisher, stored within easy reach, is also essential.

*A fireplace that's well adapted for a wood stove.*

Useful fireside tools include a poker, an ash rake, a hearth broom and shovel, an ashbucket and a canvas wood toting bag or woodbox on wheels. On the other hand, bellows, tongs and other brass-handled bits of paraphernalia usually find little use in serious woodburners' homes.

## INSURANCE

If you hold a homeowner's insurance policy, but haven't had a woodstove in the house before, you'd better contact your insurance agent. If your woodstove installation is a safe one and your insurance company is one of those responsive to changing times and the new interest in wood heat, the fact that you've added a solid fuel burner shouldn't cost you an extra penny. On the other hand, many policies contain a clause that says something to the effect that adding "a significant fire hazard" to the dwelling without notifying the company could invalidate the insurance coverage.

There is the possibility, of course, that your insurance company will cancel your coverage because of a new woodstove installation, and you'll have to look for another insurance company. If you don't report the stove, however, and a fire in your house is traced to an unreported wood heater, an insurance adjuster could possibly deny your claim. Who wants to fight an insurance company in court after the house has

*When a fireplace adapted for a woodstove has a wooden mantel, a special protective heat shield must be installed.*

*A safe stove installation*

# HEATING WITH WOOD

Some Useful Woodburning Tools: *An airtight stove with a pot of water on top to produce moisture in the air; a cast iron ash rake; a hearth broom (this one a "Double Happiness" from The People's Republic of China); a poker; a sturdy steel can and shovel for ash collection. A slate hearth protects the floor beneath the stove; a brick wall deflects heat from behind.*

burned down?

Unfortunately, the increase of interest in wood heat has resulted in a growing number of house fires traced to woodstoves. The stoves themselves are not inherently dangerous, of course; it's just that many of them have been improperly installed or are carelessly operated. Some of these installations have already taken their tolls; others are accidents waiting to happen. All these fires could be easily avoided by standard safety measures, but because they say homeowners often make error after error installing their stoves, many insurance companies aren't willing to take the risks.

This is why you should request an on-site inspection when reporting your woodstove to an insurance company. Agents are being told to watch out for dangerous hook-ups, but as one large insurance company executive noted, "if your stove is properly installed, you don't have to worry about having coverage cancelled." A top-quality stove installation coupled with an on-site inspection should avoid any insurance hassle, as well as reassuring you about the safety of the wood heater in your home. When the insurance agent arrives to inspect your home, impress him or her with your safety precautions!

# (12) How a Wood Stove Works

ONE FACT THAT separates the recent renaissance in wood heating from its counterpart in earlier times is that burning wood today can sometimes seem so *complicated*. It used to be that your grandfather would go out to the woodshed, gather a few sticks, throw them into the old parlor stove or end heater and they would give back a bit of heat. Sure, they didn't last long and there was a lot of cutting, splitting and hauling to be done, but that was part of keeping warm in the winter wasn't it?

Well, today things have changed. Stove manufacturers are now including airtight loading doors, heat baffles and smoke chambers on their products. They are talking about long flame paths and designs that "burn the gases." All in the name of efficiency. Everybody seems to like efficiency. It's got a nice, practical ring to it, and who doesn't want to get every bit of heat they can out of a piece of wood? But does anybody understand how all these things work? In a quiet moment, did you ever look at a woodstove burning merrily away and wonder "What's really going on in there?"

To begin with, wood doesn't burn. It just looks that way. What is actually happening is a three-step process called pyrolysis. When wood is heated its major components—two types of cellulose and a glue-like substance called lignin—begin to break down. In the first stage of pyrolysis, carbon dioxide and water vapor are driven off from the wood, no heat is given off until nearly all of the water vapor is gone. (This is why wet "green" wood ignites more slowly than seasoned dry wood.)

When the water vapor is almost entirely dissipated, the wood begins to give off combustible gases. This is the "ignition point" of wood. If sufficient oxygen and heat are also present, the gases will burst into flame. As long as the heat is maintained, the carbon monoxide, methane, formaldehyde, formic acid, methanol and hydrogen will continue to burn as the wood cells disintegrate.

The process also releases highly flammable tars. If droplets of water vapor are still present (as in green wood) these tars are carried off as smoke, to be deposited in the chimney as creosote. (Dry wood, containing much less water vapor, leaves most of these tars to be burned in the combustion chamber. Certain woods, notably softwoods, also contain a greater proportion of tars than other species.)

If the primary pyrolysis products are retained in the combustion chamber, they undergo further pyrolysis and combustion. (This is what is known as "burning the gases.") If they are not retained, they travel directly up the chimney. (A long flame path results in more pyrolytic reactions.)

With large amounts of heat and fuel in the stove's combustion chamber, the only practical method of controlling the rate of burning of the gases is by regulating the oxygen supply. By allowing the fire just the amount of oxygen necessary to sustain combustion at the desired level, the airtight stove permits more complete use of the fuel. However, no woodburning stove yet designed burns a hundred percent of the available fuel. Most operate in the thirty to sixty percent "efficiency" range.

Depending upon the quality of the combustion, one-half to two-thirds of the heat potential of the wood is liberated by flaming. The remaining gases are consumed by the glowing combustion of charcoal (red coals). When the supplies of carbon monoxide and hydrogen eventually fail, the flow disappears and the fire goes out, leaving only a residue of inorganic ash.

The heat from the fire heats up the surrounding mass of the woodstove's firebox, which in turn conveys that heat to the surrounding room in one of three ways: by circulation, by convection or by radiation (or a combination of the three).

Circulating heaters have an interior firebox that is surrounded by a ventilated steel cabinet with fins that direct the air current. As the hot air leaves the space between the inner and outer jacks along the routes dictated by the fins (or electric blower), cool air moves into its place to be similarly heated, and air currents are established.

Hot air emanating from a stove heats cooler air with which it comes in contact through the process of convection, which is essentially a transfer of energy that results because of the difference in density between the hotter and cooler air. It all balances out in the end, and the air maintains a fairly stable temperature during the process.

Wood heaters without circulators surrounding them rely primarily on radiation to convey their heat. The radiant stove heats the people and objects in its vicinity, which then radiate the heat themselves out to the surrounding air. As long as the stove continues to radiate heat, the relayed radiation pattern continues. When the stove stops heating, everything gets cool. Smoke chambers, found mainly on radiant-type wood heaters, provide greater radiation of the heat generated inside the firebox, because the high temperature fire exhaust is exposed to a greater surface area before it goes up the chimney, which gives it a greater chance to transfer its heat to the surrounding mass to be radiated into the room. Heat baffles have generally the same effect of forcing the exhaust to give up more of its heat before leaving the stove.

Of course, this is only one way of looking at the process of wood combustion. Even the U.S. Forest Service admits that it's only a theory. As your grandfather might have said, "That's the sun's energy coming out of the wood. It's just been stored up there for winter."

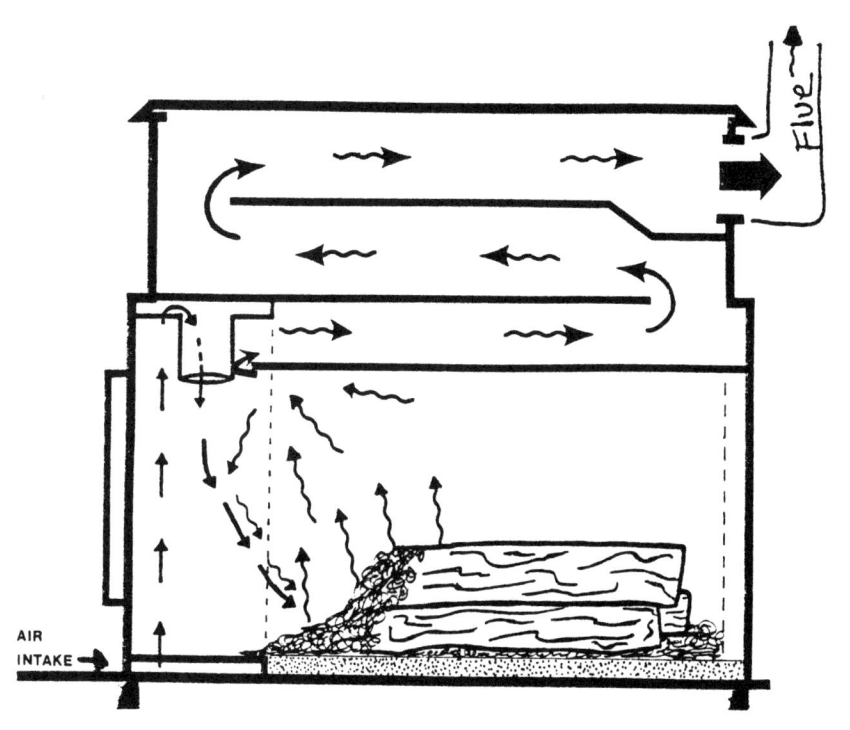

*Extensive baffling system inside a high-efficiency woodstove.*

# (13) Fireplaces

IN ONE FORM or another, the fireplace has been with us almost as long as humanity has known about the principle of combustion itself. Primitive man first captured fire for warmth, light and cooking in outdoor pits, later moving it into vented enclosures in warmer, although often smoky, caves.

Such were the beginnings of the modern fireplace. Over the eons, the fireplace has been endlessly altered, often improved, and, in its most highly developed form, made not to smoke. The device has undergone the scientific scrutiny of the eighteenth-century, nineteenth-century abandonment and a twentieth-century renaissance. By the twenty-first century, it may be all but obsolete.

Romantics, sentimentalists and backward-looking folks may chafe at such words. Indeed, the departure of the fireplace may be accompanied by a nostalgic tear from nearly all of us. There is nothing quite so romantic as an open fire, and what the fireplace loses in heat it may make up in warmth brought to the heart. Many thousands of words have been written about the aesthetic, social and even philosophical values of a bright, cheery fire, and poems and sonnets have been inspired by the leaping, sparkling flames. But in terms of cold, calculated heating efficiency, the warmth of a fireplace is largely illusory.

In the old, rambling homes of colonial times, fireplaces were the only source of winter heat, but characteristically they toasted only those in front of the hearth while leaving more distant parts of the house to the bitter cold. While performing inefficiently these fireplaces often consumed as many as thirty cords of wood in a winter—all felled, cut, split and hauled by hand. It seems safe to say that if our early ancestors had access to today's efficient wood heaters, the traditional open fireplace would have made its demise long ago.

The landmark attempt to improve the fireplace was made in 1740 by Benjamin Franklin, when he invented his "Pennsylvania Fire-Place" in the face of a firewood shortage around Philadelphia. What Franklin had produced, of course, was the first woodstove—although it looked quite different from the "Franklin stoves" of today. The Great Statesman's design was essentially a free standing, metal fireplace that projected into the room to allow for more extensive radiation of heat into the living area. He sought to minimize the enormous draft that enters the opening of the fireplace and is the mortal failing of the device as a heater. Being culturally English, however, Franklin insisted on viewing the fire as well. Ever the diplomat, he compromised with an opening that is about half the size of those found in the conventional fireplace.

In 1795, an American known as Count Rumford published a revolutionary but nearly unknown treatise on the construction of fireplaces, "Chimney Fireplaces with Proposals for Improving Them to Save Fuel, to Render Dwelling Houses More Comfortable and Salubrious, and Effectually to Prevent Chimneys from Smoking." Count Rumford's fireplace, which will be described in detail in due course, still stands today as the most efficient fireplace structure yet designed. Those who must have a fireplace unfettered by modern, efficiency-improving equipment should look to the Rumford design.

The problem with fireplaces through the ages and in the present is that they are often net heat *extractors,* drawing more heat out of the house than they produce. A roaring fire consumes lots of oxygen, producing a vacuum effect, pulling already-warmed air into the fireplace to nourish the flame, and then driving it up the chimney. While warming the people and objects nearby with radiant heat, the fireplace creates a net loss of warm air throughout the house whenever the temperature differential between the inside and outside environments is greater than 30 degrees.

A resulting inefficiency of the fireplace is that the draft created by the vacuum effect fosters a very rapid rate of wood consumption. It was this fact that first caused Benjamin Franklin to experiment with his pioneering heater, and if you are going to be chopping your own fireplace wood each winter, or even paying for it, you'll soon understand why.

There is no way effectively to control the rate of burning in

*Three devices to reform the heat-wasting fireplace.*

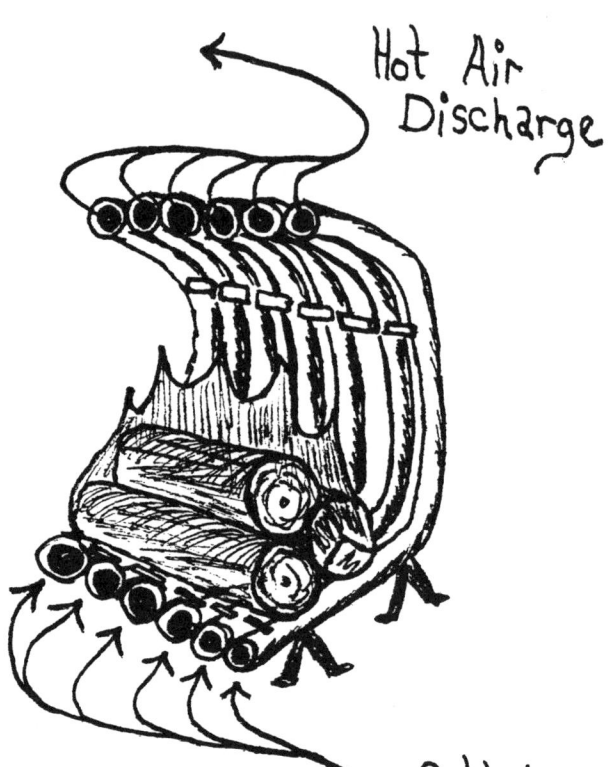

Hot Air Discharge

Cold Air Intake

a fireplace. All the air in the world is available to it, and the only way to control the fire is with the quantity and quality of wood that goes into the fireplace. This does nothing to regulate consumption, however, but simply governs the intensity of the blaze.

An additional drawback is the direction in which the fireplace deposits the bulk of the heat it generates—up the chimney. Because the fireplace hearth is located, by necessity, directly below the throat to the chimney opening, rising heat travels with the smoke along this most convenient passageway to the out-of-doors. Tests have shown that as much as ninety percent of the heat produced in a fireplace is disposed of in this manner. Put another way, of the twenty-one-odd million BTUs contained in a cord of dry fireplace wood, more than eighteen million BTUs are wasted.

In these efficiency-minded Energy Crisis years, the proliferation of woodstoves has all but been matched by a proliferation of devices to render the fireplace more efficient. Most of these gadgets are intended for use in already existing fireplaces, but some daring manufacturers are also producing equipment designed to coax energy-conscious individuals into including a fireplace in new construction. There are probably hundreds of such devices, each a variation on one of four basic themes. All of them address one or more of the problems inherent in the fireplace, but none solves these problems completely.

If you have an existing fireplace, and would like to make it more efficient, here are some of the devices you might want to consider:

## Fireplace Enclosures

These are doors that close over the entrance to the fireplace to limit the amount of air available to the fire. Without such a device, a fireplace burning in a home on a cold day would make the furnace or other main heating plant run longer than it would if the fireplace was not in use at all. These doors, usually glass or cast iron, effectively solve the heat extraction problems and give a modicum of control over the rate of wood consumption, especially if they are fitted with stove-like air vent controls, but do nothing to alleviate the heat loss problem inherent in the fireplace. In fact, by closing off the fire from the surrounding area, these devices interfere with the fireplace's normal transmission of radiant heat and thus make it almost useless as a serious heat producer.

Fireplace enclosures might best be suited to installations where a fireplace is desired for aesthetic considerations only, and the homeowner wishes to prevent additional furnace operation while enjoying the blaze.

## Tubular Fireplace Grates

These devices are designed to project more of the fireplace's heat into the room, and replace the traditional fireplace grate upon which the burning logs rest. The point of combustion is the hottest area of the fire, and it is in this region that the tubular grate gathers its heat. The tines of the grate are hollow, drawing cold air in from the bottom, and directing it around the burning logs where it is heated before being conducted by the forces of convection out of the angled upper opening and into the room. Tubular fireplace grates will not redeem the fireplace completely, but are a sensible and trouble-free addition to an existing fireplace unit, if it must be used.

## Pipes, Blowers, Ducts and Circulating Fireplaces

This related equipment is designed to capture fireplace heat and conduct it to other parts of the room or throughout the home. Although at least one unit on the market today uses a water transmission system to conduct heat through the house, most of these devices rely on the convection of heated air currents to carry warmth into the living area.

The most basic of these devices capture heat at the hearth, back wall, throat or other areas of the fireplace and project it into the nearby living space. Natural air currents may be assisted by small electric fans. The more elaborate units perform the same function, but instead of projecting heat to the surrounding room, they carry it to more remote parts of the house.

In either case, this type of equipment uses the fireplace as a furnace, and uses devices for transmitting heat that are strikingly similar to those found on the traditional furnace. But a fireplace is not an efficient woodburning furnace, and this is the fatal flaw of these systems. Pipes, blowers, ducts and circulating fireplace units recover small-to-large amounts of fireplace heat that is normally lost up the chimney, but cannot alleviate heat extraction nor restrict fuel consumption.

## Fireplace Stoves

The most sophisticated and most expensive of the fireplace-improvement appliances, these units are actually stoves in construction and design, that are intended for installation inside existing fireplaces. As such, they are the most efficient of fireplace-redeeming equipment.

The fireplace stoves are completely enclosed units with glass doors that somewhat resemble Ashley woodstoves in form and function. (Except for "Better 'n Ben's," the fireplace stove that is least like a fireplace. This unit has no glass doors and stands away from the fireplace like a woodstove, producing stove-like heat and efficiency.) Their relationship to the existing fireplace is only that they sit in the opening where a naked fire once burned—and utilize the existing chimney flue. Most fireplace stoves provide a view of the fire through glass doors surrounded by the traditional masonry fireplace, interrupted only by the sheet steel and grates that make up the circulatory heating system. Air is pulled into the fireplace stove through grates surrounding the firebox opening, heated as it passes around the firebox, and then carried into the room by convection currents.

Fireplace stoves provide the heat and efficiency one might expect of a stove; they direct heat into the room (though not as much as would be circulated by a free-standing stove), eliminate heat extraction when the doors are closed, and reduce wood consumption. Their drawback, however, besides generally not being as efficient as a standard, airtight woodstove, is that their aesthetics are as removed from a crackling, old-time fireplace as a four-color poster of a wood fire. Surrounded by sheet steel and enclosed by glass doors, the fire seems remote, burns differently, and because the heat is convected instead of radiated, warms the air in a room rather than the people and nearby objects directly, as does the open fire.

All fireplace-saving devices present a compromise of aesthetics for efficiency to varying degrees. So if aesthetic considerations are important (and isn't that the reason for having a fireplace in the first place?) it would be wise to experience these units in operation before purchasing one for your home. Definitions of beauty are individual things: One man's efficiency might be another's aesthetic Hiroshima. It's worth some investigation to be sure that your heating investment is the right one for you. Or you might just want to sublimate your thoughts about the fireplace's inefficiency, and enjoy the luxury for what it does offer. A fireplace is not a good stove, no matter what you do, it will never equal a good stove thermally.

If you are bound and determined to have a traditional fireplace, though—if woodstoves don't make it, a Franklin fireplace won't do, and glass doors and circulating units beneath a brick veneer are not your cup of tea—then you would do well to investigate the Count Rumford fireplace design.

Besides being a classic and stately structure, the Rumford is the most efficient fireplace known. It does the best possible job of projecting fireplace heat into the room, and if constructed properly, will not smoke. The venerable count's creation is the culmination of a technology that has gone steadily downhill since the end of the eighteenth century, reaching its lowest ebb since the Dark Ages in the squat, deep fireplaces of the 1950s American ranch style house. Where Rumford's invention is tall and shallow with slanted sides to project heat into the room, the contemporary fireplace is short and deep with a rectangular shape to insure that the fire's production—smoke and sparks, as well as heat—go up the flue.

## Rumford Good Heating Fireplace

### Locate Flue Inside House

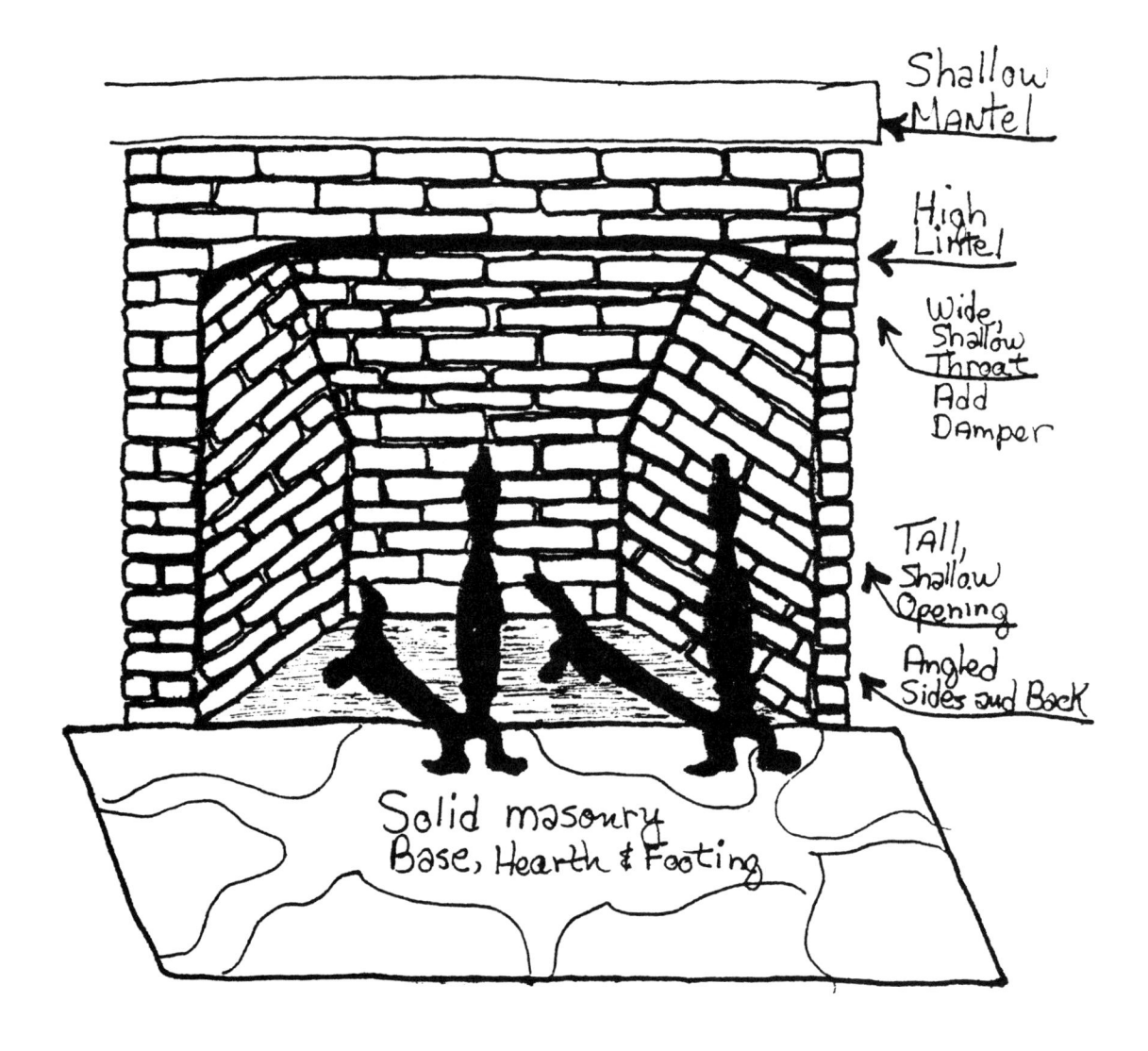

- Shallow Mantel
- High Lintel
- Wide Shallow Throat Add Damper
- Tall, Shallow Opening
- Angled Sides and Back
- Solid masonry Base, Hearth & Footing

Count of the Court of Karl Theodor, reigning King of Bavaria, Knight of the Order of the White Eagle of King George III's domain, decorated by the King of Poland, Minister Plenipotentiary to the Court of St. James's, and a colonel in the British Army, Rumford was a man of humble beginnings. Born Benjamin Thompson to a farming family in Woburn, Massachusetts, he apprenticed as a storekeeper before marrying the widow of a British colonel and coming under the benevolent aegis of John Wentworth, royal governor of New Hampshire. Later, he took his title, Rumford, from the original name of the state's capital city, Concord.

Like his peer Benjamin Franklin, Count Rumford was an inventor, statesman and genius of wide interest and ability. He completely reorganized the Bavarian army, making the soldiers better fighters by feeding them better, improved the breeding of cattle and horses, devised a method of naval signalling, studied the tensile strength of silk and the warming abilities of wool cloth, invented the first drip coffee pot and developed a method of boring cannon. He saved Munich from an invasion by Austria and France through sheer diplomacy, made major advances in the lighting of houses, the manufacture of gunpowder and the art of fireside cooking, and in the process invented a most respectable fireplace design.

Alas, however, during the height of revolutionary foment in America and throughout Europe, Count Rumford was a Tory to the last. This may explain his, and his fireplace design's, consignment to oblivion.

This is regrettable, because the Rumford fireplace deserves a niche in history. It does best what a fireplace is supposed to do: Heat. His basic discovery that heat from a fireplace is radiant heat led Rumford to a method of construction that maximized the radiation of heat into the room. Accordingly, he incorporated stone into the fireplace design instead of the cast iron that had come into use during his time, because stone is a better reflector of radiant heat. By placing the throat of the chimney flue directly over the hearth, he was able to produce a smokeless fireplace which used a smaller throat, resulting in less heat loss. Finally, he developed precise measurements for a smoke shelf in the chimney that prevented smoking and minimized the impact of the downdraft.

It is impossible to offer detailed instructions for building or altering existing fireplaces to Rumford's design here, but the basic rules of construction are quite simple. "Since it would be a miracle if smoke should not rise in a chimney (as water runs down hill)," Rumford wrote, "we only have to find out and remove these hindrances which prevent smoke from following its natural tendency to rise." In a similarly logical manner,

Count Rumford presented the method for constructing his fireplace.

The first rule is to construct the fireplace and chimney so that the center of the throat will fall precisely in the center of the hearth. At the rear of the chimney throat, construct a smoke shelf three to four inches deep, depending upon the size of the fireplace, from the rear of the throat to the back of the chimney flue. The throat itself should be three inches deep.

The sides and back of the Rumford fireplace must be angled so that as much heat as possible is directed into the room. Rumford accomplished this by building the back wall of his fireplace fifteen inches wide, and constructing the depth of the fireplace (front to back) at an equal distance. The front opening should be forty-five inches, placing the side walls of the fireplace at the correct angle. The height of the front opening may be as much as three times the depth without causing smoke problems, or forty-five inches tall.

The back wall of the fireplace should rise straight up for a distance of fifteen inches above the hearth floor, and then angle toward the front of the fireplace so that it joins the smoke shelf at the throat of the chimney. By following these measurements, it is possible to construct a fireplace which will have the correct relationship of angles for Count Rumford's design. If you're seriously considering a fireplace, consult the complete guide to Rumford fireplace construction: "The Forgotten Art of Building a Good Fireplace" by Vrest Orton, available from Yankee, Inc. Dublin, N.H. 03444.

# (14) The Chimney

WHETHER IT IS an impressive fieldstone structure, a mass of red brick, constructed of concrete blocks or modern-day asbestos and steel, the sole purpose of the chimney is to conduct smoke and gases away from the home effectively and safely. For this reason, the chimney is often the most ignored component of the total woodburning system. Though critical for safe, efficient wood heating, chimney care and maintenance often gets the short end of the stick.

Unlike stoves and fireplaces, chimneys have changed little since the first devices to conduct smoke away from the living space first appeared in medieval Europe. In the early 1900s, fire-resistant clay flue liners first appeared in smokestacks in the United States, and the most recent Great Leap Forward in chimney technology, the development of insulated steel and triple-walled prefabricated chimneys, began to appear in new houses during the 1950s.

These steel chimneys, approved by most fire codes, are less expensive to install than masonry smokestacks, (though in the case of concrete block chimneys not a great deal less). Stone, brick or block chimneys may be more costly to build, unless you do the labor yourself, but generally, they are more durable. The mortar between the masonry units will deteriorate with time, however, and must be repaired to maintain a serviceable chimney. This repointing operation simply involves chipping away the loose, crumbling mortar and pushing new mortar into the cracks. Check the condition of your chimney.

[104] HEATING WITH WOOD

*A beautiful chimney—but it squanders heat to the outside.*

# THE CHIMNEY

The masonry chimney has one additional advantage: It helps protect the house from the extremes of heat and cold, by absorbing excess heat while the fire's burning bright and slowly radiating it later when the house begins to cool—unless the chimney is built outside the living area, where it never should be.

Each heating season, before putting an old chimney into service, a thorough inspection is in order. Don't neglect this important step, or you may join the more than 40,000 Americans who reported chimney fires in their houses last year.

Stone and brick chimneys found in a great many older houses have no flue liners, so masonry cracks there are certain to pose severe fire hazards. Cracks should be repaired, of

*Better use of a fieldstone chimney: absorbing heat when indoor temperatures rise, then releasing the heat into the living space to soften temperature swings after the air begins to cool.*

course, but if you're thinking of reviving an old chimney, you should also outfit it with clay flue tiles or a steel liner. This is not as difficult a job as it might seem. It can be expensive and somewhat time-consuming, but not nearly as much of either as building a new house.

The best method of locating leaks in a masonry chimney is to build a small, smoky fire in your stove or fireplace, and then cover the top of the chimney with a wet blanket or bag. The trapped smoke will seek out every available exit, so while you're on the rooftop, a helper can pinpoint every crack in the chimney by watching for the smoke and then circling the spot with a piece of chalk. A heavy patch of rich mortar will usually repair the damage.

*A flue liner that makes a faulty chimney serviceable.*

The ideal masonry chimney consists of a sound, tightly mortared exterior enclosing an inner chimney of flue liners that extend from the top to the bottom of the unit. The clay tile liners should be mortared together and, ideally, an air space should exist between the liners and the outer masonry shell. To those folks who have never experienced the driving heat and menacing roar of a chimney fire in their own homes, such a structure may seem exessively elaborate.

But as a friend who has traced the histories of many old houses once asked, "Ever wonder why there aren't more old houses around today? They didn't just deteriorate or fade away. Most of them burned to the ground."

Better safe than homeless, I always say.

If you're building a new chimney, also be sure to place it on a generous concrete foundation, and add an ash door at the bottom of the flue for easier chimney cleaning. If you're in doubt about how many woodburning appliances you may be using, it might be a good idea to add an extra flue to the structure right at the start. No more than one woodburning stove should be connected to an 8″ × 8″ chimney flue (eight inch diameter, for prefabricated smokestacks.) An 8″ × 12″ flue can vent two efficient woodstoves, one Franklin stove, or one small fireplace. Connecting more than two heating devices to a single flue is possible but ill-advised; this practice can lead to flue gases being drawn back into the house, poor draft and excessive creosote build-up. Woodstoves and oil or gas furnaces should never be vented into the same smoke passage.

Common chimney problems include poor draft and clogging. An insufficient draft may result in a weak fire or puffs of smoke entering the house, and may be caused by a chimney that is too short, or by too many heaters on one flue. Chimneys should project two to three feet higher than the ridge line of the roof. Even the most stubborn puff-back problems can usually be cured by adding more height to the chimney, by topping it with a rotating wind cap to increase the draft, or by cleaning the chimney.

Next to a poorly constructed chimney, the greatest danger to woodburning folks and the homes that shelter them is a dirty chimney. Clogged flues may be caused by anything from fallen masonry to a bird's nest, but the most common cause of a dirty chimney is creosote deposits. Creosote is a by-product of incomplete combustion, and since neither wood nor any other fuel burns completely, some build-up is certain to occur. However, using only dry hardwood as woodstove fuel will minimize these deposits dramatically.

Extremely efficient woodstoves release less heat to the chimney, resulting in cooler stack temperatures and the potential for greater creosote build-up. The more efficient your woodstove, therefore, the more frequent your chimney cleaning chores should be. Creosote build-up can also be minimized by burning quick, hot fires rather than slow, smouldering ones if your lifestyle can handle it, by avoiding long horizontal runs in the stovepipe between the wood heater and the chimney, and by capping the chimney to keep moisture out and heat in.

The arrival of each new heating season, as well as excessive smoking when you open the woodstove door during the season, are signs that it is time to clean the creosote out of the chimney.

Creosote is long-burning and quite flammable. It will build up layer upon layer in the chimney and stovepipe, until flue temperatures hot enough to ignite it set the sticky mass ablaze all at once. A large quantity of creosote can create a chimney fire hot enough to send flaming balls shooting from the roof and severe enough to damage both the chimney and the house.

On the other hand, frequent burning of *small* deposits of creosote inside the chimney is often unnoticeable and will serve to keep the chimney reasonably clean. Unfortunately, this is usually difficult to accomplish.

A more common—and reliable—method of keeping the chimney free of creosote build-up is mechanical cleaning. Scraping creosote from the sides of the chimney lining and stovepipes can be done in a variety of ways, and inventive woodburners are constantly devising new methods on their own. However, chimney cleaning is not a job for every do-it-yourselfer. To work at the chimney tops you must have a bit of steeplejack in you and the steeplejack's ability to avoid serious falls.

If you do decide to tackle the chimney-cleaning chores yourself, the simplicity of the task should prove encouraging. The first step is to close the stove dampers and tightly cover the opening of the fireplace, to prevent loosened soot from falling into the house. The next step—tools in hand—is to mount the roof to the chimney. Some brave souls accomplish this by throwing a lasso around the chimney, and pulling themselves up like a mountain climber, but a tall, well-placed, sturdy ladder is recommended. Be careful! Tie a safety rope around your waist and anchor it to the chimney if the masonry is solid. (Older chimneys, especially, may not be able to take the stress. Test yours first before relying on it for support.)

Once you are on the roof, the job becomes downright easy. Soot is bound to blow back into your face, though, even if you do stand on the windward side of the chimney—which is recommended—so eye protection and a face mask to filter out creosote dust are a must.

[110] HEATING WITH WOOD

The chimney can be cleaned out with a standard chimney sweeping brush or with a number of homemade devices. One of the most effective of these is a burlap bag stuffed with enough crumpled chicken wire to fit snugly into the flue. A rope is tied to each end of the bag. (Each rope should be at least as long as the entire chimney is tall.) While one person stands on the roof holding the end of one rope, a helper holds the other rope, which has been passed through the cleanout door at the bottom of the chimney or through the lowest flue opening. The bag is then pulled up and down the length of the chimney repeatedly, until all the built-up deposits have been scraped away, and the tinkle of falling creosote chunks ceases. Shine a flashlight down the flue to be sure the cleaning job is complete.

*One type of chimney brush.*

THE CHIMNEY [111]

*Another type of chimney brush.*

An inverted Christmas tree or other homemade scraping device can be substituted for the wire-filled bag, but doing things like banging old tire chains around inside the flue to loosen creosote is likely to crack the tiles and generally cause more problems than it solves. If in doubt about a device for cleaning your chimney, a safe bet is to buy a custom-sized chimney brush from your woodstove dealer.

After cleaning, it is a simple matter to sweep the soot out of the cleanout door or the fireplace. (If you have a fireplace, remember to sweep off the smoke shelf just above the "throat" or flue opening.) Clean out the stovepipe with a brush, stick or poker, too, and save all the sweepings for lawn or garden fertilizer.

If you burn wood regularly and neglect these chores, be prepared for an unpleasant surprise. At best, a chimney fire roaring out of control can damage bricks and tiles while weakening mortar, while scaring some caution and common sense into the folks at home. At worst, the leaping flames can reach through their masonry prison and engulf the entire house. Though such a fire can be a frightening experience, fire marshals, masons and woodstove installers agree that chimney fires need not occur. With a responsible woodburning attitude and frequent maintenance, any family can benefit from wood heat with the same confidence they would place in electric baseboard heating.

If a chimney fire does get started in your home, don't panic. Throw a handful of coarse salt on the fire to douse the flames, close down the draft controls, and call the fire department. Clay flue liners will crack if cold water is poured on them during a chimney fire, so don't don't spray water directly into the chimney unless absolutely necessary—as a last-ditch effort to save the house, for example.

Here's a final caution: In more ways than one, chimney cleaning can be hazardous to your health. People have fallen off roofs since chimney cleaning first began, but as early as the eighteenth century, British scientists noted a high rate of cancer of the scrotum among chimneysweeps. More recently, coal tar residues have been suspected of causing other types of skin cancer. The simple defense: Avoid breathing ash dust and scrub *thoroughly* after finishing the chimney-cleaning chores. A top-to-toe shower is recommended.

If all of this seems too much to tackle on your own, you can always call your local chimneysweep. Some folks might guess that chimneysweeps went out with the nineteenth century, but the trade has experienced a comeback along with the renewed interest in heating with wood. In fact, the medieval profession has recently spawned a medieval trade association: America's first Chimney Sweep Guild.

Many 'sweeps work in modern, protective headgear and orthodox work clothes, but members of the Guild often appear on the job clothed in the traditional chimneysweep's garb—black top hat and tails. The 'sweeps are also reviving old customs, like appearing at weddings in full dress. "Remember, it's good luck for a bride to be kissed by a chimneysweep," one practitioner advises.

For serious woodburners, though, the 'sweeps bring good luck just by tending to their chores and helping to prevent chimney fires.

To contact a chimneysweep in your area, get in touch with the Chimney Sweep Guild, c/o Kristia Associates, P.O. Box 1118, Portland, Maine 04104, or check the Yellow Pages.

# (15) The Fire

PERHAPS YOU'RE A newcomer to the world of wood heat, following your disenchantment with petroleum-based heating to a sensible alternative, and your instincts (and the information contained in this book) toward providing your own warmth. Perhaps you've read and pondered, planned and reflected some more, then taken the plunge and purchased a good woodstove. Right now, that stove may be standing before you, safely connected with loading door ajar, waiting only for that first flame to bring it to life.

If so, savor this moment. For you are about to re-establish your roots, to reach back into your ancestry to involve yourself with Fire Essential to Life in an immediate, personal way. Fire is a critical ingredient in the evolution of the human species. Learning to harness its energy was the prime factor that allowed primitive humans to move out of the earth's equatorial regions to inhabit the northern and southern areas. Once again, as the twenty-first century draws near, the basic, honest wood fire can provide the warmth needed to survive future frigid North American winters.

On the other hand, it may be that you have tried your hand at firebuilding before: touching off a blaze in the fireplace, firing up a cranky antique stove resurrected from the basement during the Original Energy Crisis of 1973, or feeding a campfire. If you haven't been able to get your wood to light with any dependability, if you've had smoke pouring back into the living room, if your fire has flamed, fizzled, glowed and died long before its time was due, if you sometimes find yourself red-faced and ash-covered, hyperventilating yourself dizzy trying to blow some life into the flames, then you too should take a moment to pause and celebrate. Read these pages, practice a little, and your fire huffing and puffing days will be over. Here's how to build a safe, reliable heating fire.

As if by magic, one tiny flame at the tip of a candle can create another tiny flame of the same size on another candle without giving anything of itself away. Taking this example to its extremes, the same tiny flame can create a raging inferno, where glass melts and temperatures reach thousands of degrees, if the three necessary componets of fire are present. All it takes is combustion temperature (heat), oxygen (air), and fuel (wood, of course).

The key to good fire-building is establishing an incendiary pyramid, and the proper procession of fuel types. The most quickly igniting, hottest burning form of wood is thin layers of fiber, held together in large quantities, with plenty of airspace around them: i.e., newspaper. A newspaper fire will light small pieces of dry kindling (this is the place for softwood, like pine branches), and the kindling, in turn, will light your small logs, and eventually, your serious fuelwood. Build in a pyramid shape, with plenty of open passageways for air to pass between, among and around the solid fuel, and you'll always have a good fire.

Those are the principles, but there are also some practical considerations involved. One is the moisture content of your fuel. Another is temperature. If your wood (especially kindling and small logs) is freshly cut and green, frozen or soaking wet, the chances of smoothly touching off a stable fire with it are nil. After stocking their stoves with this kind of stuff, some frustrated fire technicians then resort to dumping materials like gasoline, kerosene or charcoal lighter fluid on the wet or frozen wood. Even if they survive their folly intact (many find themselves *sans* hair, eyebrows or moustaches), these poor souls still find themselves without a fire, because after the highly refined fuel has burned away, the unfit wood still does nothing but smoulder. (I know; I tried it once myself.)

Don't use liquid fire starters. They're as dangerous as they are unnecessary. On the other hand, don't try to start a fire with unsuitable wood fuel either. It's as unsettling as it is inefficient. Don't fight with your woodburner. Use only dry, well-seasoned firewood, protected from the elements, and bring it in the night before, so it can warm to room temperature before you try to light it. (This makes the job a lot easier.)

Some people say there are many ways to start a wood fire,

but as far as methods for "laying a proper fire" go, to my knowledge there is only one. This method originated, so far as I can tell, in England—often called "the most civilized of countries"—generations before modern energy crisis consciousness hit twentieth-century America. It was passed on to me many years ago by a tall, cultured, aristocratic but kind British matron who knew how to build a fire. I've tried many other firebuilding techniques, and watched demonstrations of several more, but despite the fact that most of them resulted in a dancing wall of yellow flame, eventually, I've never seen a firebuilding method that yielded fire more quickly than this one.

To lay a proper fire, ball up six sheets of a full-size newspaper into separate wads, each about the size of a baesball. Arrange them in two rows, like a half-dozen eggs, under the grate of a fireplace or combination heater or on the floor of the firebox in a typical box stove. On the top of the newspaper lay a generous bed of finger-size kindling, perhaps twenty pieces for a sure start. *Be sure to lay all the kindling in the same direction* for a solid flame. Good kindling materials include dry shingles, split pieces of old dry lumber, or thin splits of pine, spruce, fir, or other softwood. (Hardwood kindling is much harder to ignite.) You may have noticed that dead pine branches still clinging to the lower part of the tree remain brittle and dry even after a snowstorm. If a pine grove is handy, this is a good source of kindling wood in emergencies—or throughout the year.

After the kindling is laid, place not one, not two, but three, dry, small or split logs about two to three inches in diameter, atop the pile. Burning wood needs other burning wood alongside it to maintain combustion temperature, which is why you can almost never light a single log. Stack your wood in a triangular configuration over the kindling—two logs on the bottom, one laying between them on top, all in the same direction as the kindling for good heat transfer—and the flames will lick up through the opening between the bottom two logs and curl around the top log, setting them all on fire. (Once again, be sure there are enough air spaces between the kindling and each of the logs to allow good air flow. Oxygen is as crucial to the fire as the wood itself.)

At this point, you're ready to strike the match, and if you've followed these instructions carefully, a single match is all you should need to light a fire. Open all dampers and draft regulators (most airtight stoves have two) all the way to create a good air flow through the stove and up the chimney, and light up!

As the fire crackles to life, and then begins wrapping itself around the logs, start closing down the direct air regulator on your airtight stove, and if you have one on a stove or fireplace, begin easing down on the damper to keep too much heat from escaping up the flue. Use the preheated air regulator (if your stove has one) to govern the intensity of the fire. Once the three small logs are ignited and burning well, you can fill the stove's firebox to capacity with larger pieces of heavier woods like oak and maple, or stack more wood on the hearth or in the open Franklin as you desire. Maintain the pyramid.

By feeding large chunks of dry hardwood to an airtight stove, and restricting air intake to the amount needed to sustain a comfortable fire, you can keep this same fire burning all winter long. Many efficient stoves on the market today will hold a

*Load the firebox with six wads of crumpled newspaper, arranged like a half-dozen eggs.*

[118] HEATING WITH WOOD

fire sufficient to warm a house for eight, ten or even twelve hours on a single charge of wood. Still, to avoid waking up to a cold house some morning, or to prevent coming home to icicles dripping from the faucets, you have to use that burn period to advantage.

If you're planning to use wood as your primary heat source, establish a burning and loading schedule that will carry the stove through long periods of inattention, such as when you are off at work, or while you are asleep. Try to begin a new burning cycle before you go to work, or just before bedtime, when all that remains in the firebox is a bed of hot, red coals. This way, you'll be able to fit enough fuel into the woodburner to carry it through the day—or night. (Otherwise, you'll have to "sleepwalk"—get up in the middle of the night, stoke the

*Cover the neswpaper with a generous bed of kindling. All sticks must be running the same way.*

stove, or worse still, stumble around for matches and kindling, then go back to bed.)

Three, eight hour burning cycles per day is ideal, if your fuel is good enough and your stove is efficient enough to handle it. In any case, try not to recharge a stove that's merrily perking along until it's nearly empty. Opening the door frequently only disrupts the heat flow of an efficiently burning stove.

If you're planning to use wood heat occasionally, as a supplement to solar heating or another type of fuel, stock your wood supply so that you'll have plenty of newspaper, kindling and small logs on hand to start the blaze each time you want warmth from a fire. For quick heating, stay away from the oaks and other heavy woods, and favor the light woods like

*Add a pyramid of well-dried firewood.*

# [120] HEATING WITH WOOD

birches, and poplar. These woods will yield a bright, cheery, warming fire if given enough air to burn hot and fast, making a pleasant evening blaze. They leave little in the way of coals, though, so if you want to bank the fire for resurrection the next day, use more substantial firewood, and after the fire is out, cover the glowing red coals with ashes, or ten to fifteen sheets of glossy magazine pages. These substances will hold the heat inside the pocket of coals while restricting air intake, keeping combustion to a minimum. Then, like the old cooks on the open range, you can stick some chunks of pine into the coals in the morning to get the coffee steaming again.

There's lots of heat value in much of the trash that passes through American homes today, and if you take the time to

*Touch off the paper with a match, and soon your fire will be blazing.*

sort out the combustibles and to burn them *carefully and safely,* these BTUs can be turned to your advantage. Please note, however, that if you burn a stove full of paper with all dampers open, you run the real risk of cracking or warping your stove, ruining your stovepipe, damaging your chimney, and who knows, maybe even burning down your house. *Be careful!* If you turn around, and notice that you've let the hot, fast-burning trash fuel turn you stove cherry red, you've blown it.

While we're in a cautionary vein, let's also be sure to burn only paper and cardboard products. Putting plastics in the stove will not only result in a sticky residue, it will also put some dangerous, possibly cancer-causing materials in the smoke coming out of your chimney and render your wood ashes unfit for fertilizing the garden. Metal objects will do little harm in the stove, but will do no good either.

Okay, let's say you've established a separate waste bin in the family area exclusively for waste paper products, to be fed to the living room stove. On a chill spring morning, or when the frost first begins to fall on the pumpkin in early autumn but it's still too early to fire up the stove for an all-day or all-night burn, turn to the waste bin to save on cordwood supplies.

Roll newspapers, pizza boxes and other pieces of burnable trash into log-like cylinders, stack them in the stove, light it, close the door, and as soon as the charge begins to burn, close down the draft regulators. Let the first charge of waste burn completely before opening the stove door again, to prevent flaming sheets of paper from falling out of the stove onto the floor. Then load again, if desired, until the room is warm, and the chill is out of your bones.

When you are not using the woodstove or fireplace, always be sure the drafts and dampers are closed tight, to prevent hot air from escaping your house into the Great Outdoors in winter, and to stop bugs from drifting down the chimney into the house during the summertime. (An open chimney flue *can* be useful in drawing heat out of the house in summer, of course, as long as the opening is screened to prevent insect invasions.)

If you are touching off a fireplace or open combination heater, warming the flue before lighting the fire will prevent a good deal of back-smoking into the living area. A cold flue won't draw, as they say, so smoke from an open fire will often puff back into the room until a warm updraft develops in the chimney. To avoid this, preheat the flue by holding a half-folded sheet of newspaper under the chimney throat and light it. After it's burned almost to your fingers, use the remaining flames to touch off your blaze, and the smoke should begin rising immediately. Should your stove or fireplace continue to smoke check for obstructions in the chimney. (See Chapter 14.)

An untended ash build-up can also restrict normal burning, by clogging grates, which cuts off the fire's oxygen supply, or by taking up so much room in the firebox that normal circulation patterns are disrupted. If your stove is equipped with an ash pan, make emptying it a regular routine, certainly before it's filled to the brim. The chore is much easier and less messy this way. If your stove doesn't have an ash pan, shovel out the firebox as soon as the ashes reach the door opening.

Before lighting any stove for the first time, be sure to read *every word* of the manufacturer's directions carefully. With most box stoves, it's best to line the bottom of the stove with an inch or two of sand, or even firebrick with sand poured between the joints. This protects the floor of the stove from the intense heat concentrated in the hot coals beneath the burning wood.

After your stove is suitably lined and you've "laid a proper fire," touch off the blaze and watch all joints and seams in the stove, stovepipe and chimney for cracks or leaks. If you do see smoke wafting out anywhere, remedy the problem with furnace cement or a good masonry patch before further burning.

When you first fire a new cast iron stove, an acid condensate may form in the firebox and along the smoke path. If it leaks out, this substance could etch the iron or stain enamel permanently, so watch the stove for its first couple of hours of burning and wipe off any condensate that may appear. After the stove has been thoroughly heated once, there will be no more of this acid condensation.

Most importantly, make your first three or four fires moderate blazes, not flaming infernos, which can ruin an unseasoned stove. Never do anything that will cause sudden, sharp changes in the temperature of a stove, such as throwing cold water on a hot stove, or firing a cold stove full of pine kindling with both drafts open.

Keep your stove blacked if necessary to avoid rusting. It's also a good idea to rub down the interior and exterior metal parts with light oil, if the stove will not be used for long periods of time.

If your fire is burning too quickly and too hot or too cool and slow, check your draft regulators. The more air supplied to the fire, the quicker and hotter the fire will burn. Also consider your choice of firewoods. From pine to hophornbeam, there's a use for every species in the woodlot, but each type of firewood has its place. A couple of slabs of pine will get a charge of knotty oaks logs a-cookin' real soon, but a piece of oak buried in a firebox full of pine slabs will do little to temper the resulting fierce, brief fire. Be judicious in your fuel selection, and mix for the specific heating goal you have in mind. In time, with practice, you'll cultivate the art of woodburning.

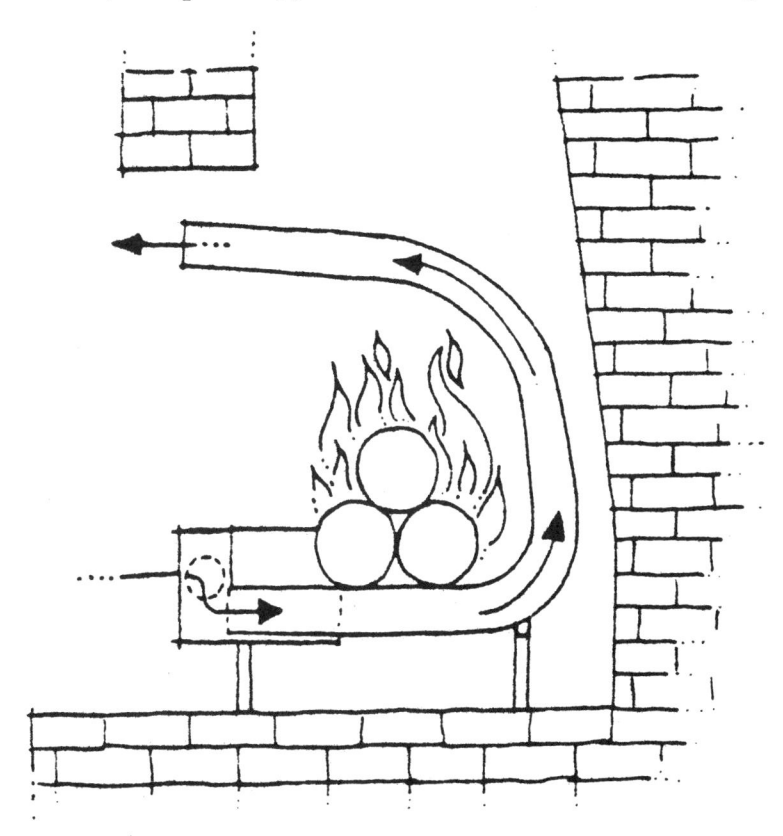

# (16) Cooking on a Wood Stove

EVEN BEFORE WE STEPPED out of the car in front of the century-old farmhouse, the smell of a woodstove burning told us we were at the right place. A cowbell clanged softly in the distance as we walked up the maple-lined driveway to meet Florence Geddes (affectionately known as "Gram"), a New Hampshire native who has never cooked with gas or electricity in all of her eighty years.

We went to see Gram Geddes sometime ago, because we were then considering buying a kitchen woodstove for our budding homestead, but were somewhat reluctant, owing to the popular belief that cooking on a woodstove is difficult, time-consuming and oftentimes a chancey endeavor at best. By the time we had left, Gram Geddes had convinced us otherwise.

The woodburning cookstove is a creation unto itself, whether it's a traditional nickel-and-iron Queen Atlantic, a "modern enameled model, or a new Scandanavian design. Producer of the finest of breads, pies and roasts under the guidance of an expert master or mistress, the kitchen stove can also turn ornery, sullen and well-nigh impossible to light. How hard would it be to master the fine art of woodstove cookery? We hoped this octogenarian expert could tell us.

Gram Geddes was standing over her yellow enamel Sears, Roebuck kitchen cookstove when we arrived, a simple piece of equipment she and her late husband bought second-hand about thirty years ago. "I couldn't think of having breakfast without it," she mused as she peeked at a loaf of bread baking in the oven. It was dinnertime and Gram was putting the woodstove through its paces.

"You can cook a good many things on here at one time," she said. There were two teapots steaming away on the stove to provide "cooking water," an old-fashioned pot roast simmering in another corner, bread in the oven, and several dozen cookies and a sweat bread that had just come from the oven on a table nearby.

The woodstove also preheats Gram Geddes' water. A pipe passes through the firebox in a loop, carrying the cold water supply to the stove to be warmed before it enters the electric hot water heater. "We don't use so much electricity that way," she explained. "We used to heat all our water in the stove, but as the family grew all the hot water got used up for baths, so that we had to put in the electric unit."

In her unvarnished kitchen, Gram Geddes tended the food constantly as she spoke to us. She made it seem as natural and uncomplicated as making a piece of toast. We asked her to tell us about her secrets.

"You go about cooking on them like you do on any other thing, oil, gas or electric," she said. "The only thing different is that you can't turn the heat down on a woodstove like you can with automatic heat. A woodstove holds the heat a long time. If you think the oven is going to scorch your bread, and you turn down the flame, it's going to scorch your bread anyway.

"You need a good fire for the oven. My mother used to tell us to build a fire for bread-baking, and then put an arm in the oven and count to twenty. If you couldn't keep it in that long, the stove was too hot. "Course I don't bother with it now—a look at the fire and a feel in the oven and I call it a job. There's a thermometer on the oven door, but it's practically useless—I think most of them are. I've got one hanging inside the oven, too. It tells the temperature all right, but I don't pay much attention to it."

Undoubtedly, the process seemed simple to a woman who had been cooking on woodstoves for a farm family for most of her adult life, but for the benefit of neophytes we asked her to explain her method. "First thing you do is check on your fire," she explained in solid schoolmarm fashion. "You've got to have a good fire before you can do a thing. When the oven is preheated, begin your baking first. Then you can start cooking on the top of the stove."

Gram Geddes stopped to pepper her pot roast, a traditional way on an old-fashioned woodstove. "I'm a firm believer in a long time, slow cooking," she said. "I believe it makes the food

[126] HEATING WITH WOOD

*Gram Geddes: eighty years young and always cooked with wood.*

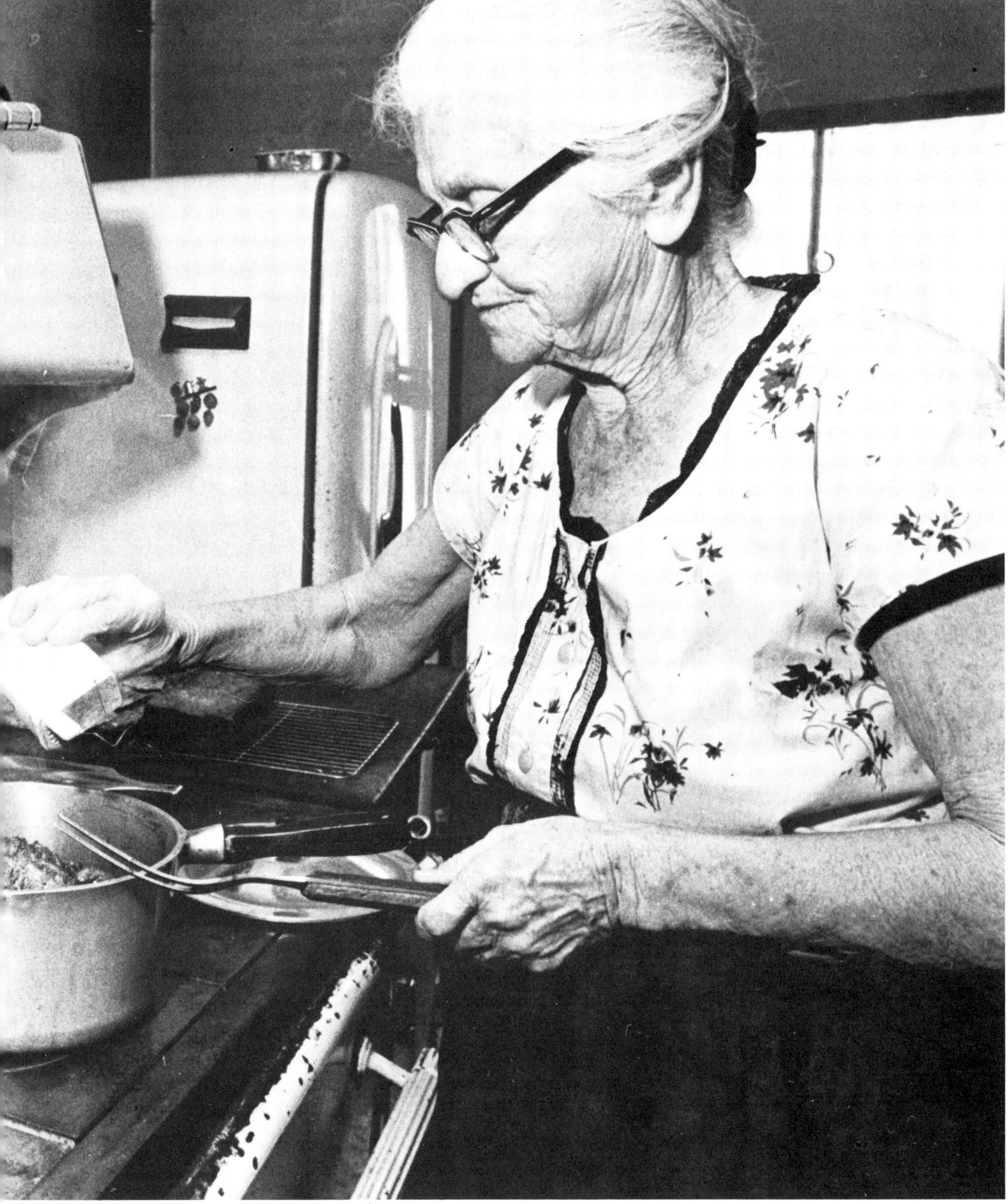

taste better. I don't know what I'd think of dinner cooked in one of those electronic ovens. What do you call them? Oh yes, microwave.

"Well, it takes me a half hour to boil water on this cookstove from the time I kindle the fire," she said, "with good wood that is. And I believe I can heat an oven as fast as you can heat one of those small electric ovens." The right type of wood, we learned is indispensable to the smooth operation of a wood cookstove.

"If you're cooking with a woodstove," she said between peeks at the rapidly-rising loaf in the oven, "one thing you've got to have is good kindling. I use wood 'splinters,' left over from the splitting logs outside. 'Course wood shingles can't be beat if you've just put new shingles on a barn and have the old ones left over, and pine cones are terrific. They make a roaring good fire.

"The best way to get a long fire is to get a good bed of coals going, and put two substantial pieces of wood on them. I can can keep a fire overnight in this way, so that when I come out in the morning there's a little heat in the kitchen. I can get the water boiling faster and I don't have to strike up a fire from scratch."

Gram Geddes called rock maple her favorite cooking wood ("It's really too good to burn in the cookstove," she said with a smile), although she rated white ash "very good."

"In fact, if you must burn green wood," she advised, "use ash. If it's dried one week, it burns almost like dry wood. Beech is also a very good wood for cooking. Oak is what you use for a long, low fire because it burns so slow, and grey birch isn't much good except for a summer fire. It burns hot and quick, and leaves no hot ashes, only pink coals."

Whatever type of wood you burn, though a traditional woodstove—neither airtight nor terribly efficient—will eat plenty of it.

Gram Geddes slipped the browning loaf out of the oven and turned it around with a deft motion. "Most folks think you have to turn whatever you're baking around in the oven," she said, and we agreed. "It's not necessary if you've got a proper oven door, but mine doesn't close too well anymore. This is just a makeshift latch that holds it shut."

A great advantage to the wood cookstove is that different regions of the cooking area heat to different temperatures. The front of the stove is the hottest area, and it is here that most of the cooking is done. Gram puts her pots on the front "to get things going," and then sets them back a bit to simmer. Teapots will steam merrily at the rear of the stove, and food will keep warm over the water storage tank or in the top shelf warming ovens. To get a red hot fire, you can fill the top-loading firebox right to the brim and get added heat to cast iron pans by removing the round lids from the top of the stove and placing the pans right over the flames.

A woodstove is also without peer for warming cold kittens who cuddle behind it, drying winter mittens and boots, incubating struggling newborn chicks, taking the edge of green stovewood and drying out the hand towels that Gram hangs behind it.

Of course, a woodstove does require some maintenance electric heat doesn't, and that's cleaning the firebox. It's a rare day that passes without Gram Geddes lighting the fire in her stove, so she cleans the cookstove—herself—twice a year. "I've got barrels of ashes by springtime," she said, "They're terrific for the garden, or for spreading on the lawn or the flower beds or just anywhere you want things to grow."

It may seem like quite a bit of trouble to tend a woodstove for each family meal, but Gram Geddes wouldn't have it any other way. "I've known McGees, Crawford, Glenwood, Kalamazoo, Home Comfort—which we disinherited when we got this one—and they all seemed to do the trick. I've cooked, boiled, baked, steamed, and canned on them, and they're not too awfully bad about heating the kitchen when it's cold outside, either." Just then, Gram's timer began to buzz, and a perfectly formed, golden brown loaf of bread came from the oven.

There was a time, not long ago, when a casual observer might have suggested that all wood cookstoves were just about the same. Dedicated partisans would vigorously defend their Queen Atlantic, Glenwood, Kalamazoo or other favorite. Still, one might have said (as Spiro Agnew did in an even less appropriate context), "If you've seen one, you've seen them all."

Of late, however, there are several breeds of newcomers elbowing their way into the kitchen space of woodburning America. Some are wood/oil, wood/gas, or even wood/electricity combinations, while others, the newest generations of cookstoves, are usually of Scandinavian design and are smaller, airtight and therefore more efficient. They often possess an aesthetic character that is quite different from the massive bulk of more traditional kitchen ranges.

It takes a somewhat different set of skills to skipper one of these modern wood cookstoves, so as the September leaves turned to crimson and gold we set out to find someone who has mastered the newfound art.

It didn't take too much searching before we found Arthur and Judy Davidson's home on a backwoods road in the hills of Northfield, New Hampshire. A thin waft of smoke floating from the fieldstone chimney told us we were at the right place.

Inside the sunlit kitchen that Arthur himself had built ten years before, Judy was stoking up a dinner fire. A bright blue enameled Lange cookstove stood before her, resplendent in its adornment of a gleaming stainless steel kettle and stewpot. Dinner that night was to be a garden stew, and as she sliced a tomato from a basket of homegrown vegtables, Judy accounted for her experience with a modern cookstove.

"We've had all sorts of stoves in here," she said. "First there was a Riteway, then an Ashley, then got a small cookstove—too small, we found out later, even to keep the kitchen warm through the winter. Finally, we got a big old Home Comfort cookstove, and that served us well for quite awhile.

"It was only last year that we had an opportunity to get a Lange cookstove," Judy continued. "I told my husband I'd go along with it, but I didn't want to sell the old stove until I had a chance to try the new one out. Well, the Home Comfort stood on the porch for about three weeks. We soon decided that we didn't want the old stove anymore."

For most of the year, the Davidsons rely on their woodburning cookstove to feed themselves and their three children. Only in the hottest months of the summer do they turn to the tiny electric range nearby.

"With the old cookstoves you start cooking a meal and pretty soon the temperature is up to about 90 degrees in the kitchen," Judy said. "Then, a couple of hours later, it's cold in the house again. This stove is cool to the touch (she rested her hand on the protective brass railing) so it's comfortable to cook with, but it will hold a fire overnight. In the morning you just add some wood, open the drafts, and you have your teapot singing in no time.

Users of traditional woodstoves often devise elaborate stoking and banking routines to keep the precious glow in the firebox overnight against the early morning chill, but with the newly engineered cookstoves, maintaining a warm winter kitchen is no trick at all. "Last year we just about ran out of wood," Judy recalled. "All we had left was poplar. But this stove has a good big firebox and even a few sticks of poplar held the fire until morning."

The Lange has two traditional stove lids on the cooking surface that can be removed to put a pan in direct contact with the heat. "Europeans often like to cook that way," Arthur said. "In fact, they say that a blackened pot cooks better." The Davidson's stove is not fed wood through the top, as is the case with old-fashioned cookstoves, but through a front-loading door similar to those found on other Scandinavian woodburners.

"The entire top of the stove is a cooking area," Judy explained, "although there are different tempearturs in different regions. The center of the stovetop is the hottest. This is where you put your teapot when it's full of cold water in the morning. The area over the firebox is hot enough for cooking and the right side will keep your food warm.

"The oven is all cast iron, so it distributes the heat a bit differently than ovens that are sheathed with sheet metal inside," Judy continued. "Generally it heats pretty evenly, but your bread might tend to get a bit browner on one side than the other. If you have a roaring fire, the loaves might need to be turned.

The oven in most of the Scandinavian cookstoves appears

*Judy Davidson skippering a modern cookstove.*

to be quite small. "But it is well-proportioned, so it gives you a lot of cooking room" said Judy. "I find that I can bake three sheets of cookies at one time in it, and it is deep enough to hold three loaves of bread." A large enamled pan custom-fit to the oven comes with the Lange cookstove, "large enough for a good big lasagna," she said with a smile.

"And pancakes!" The smile broadened to a whet-the-appetite reflection. "We put this cast iron griddle right on top of the stove. You sure don't have to wait long in the morning for pancakes."

In the past, cookstoves have been judged primarily on the way they cook and any surplus heat was considered an added bonus or perhaps penalty, depending upon your distance from the stove. Yet the Davidsons consider their cookstove's heating abilities to be one of its greatest assets.

"In the winter we heat most of the house with it," Judy said. "And the wonderful thing is that the temperature doesn't vary very much. You can cook on it for hours and the temperature might rise from 70 to 75 degrees. At night it might drop to 55 —instead of 35 like it did before. With the heat that comes from the woodstove, we depend pretty much on the sunshine to warm the rest of the house during the day."

Maintenance of the cookstove consists primarily of emptying the ashes regularly, but the efficiency of Scandinavian stoves produces relatively little residue, so even this chore isn't done too often. "Every cook has to get to know a stove for awhile," Judy concluded, "but once you get the knack of it, you can go from a cold stove to the breakfast table in less than half an hour. It's really not hard at all."

# (17) Cookstove Recipes

It's mid-January, and you've been out cutting firewood all day. Now the sun is setting and you're *hungry*—hungry for food that will please your palate, satiate that growling stomach and still cook in a reasonable amount of time.

Well, if you're looking in the direction of your wood cookstove, you're heading in the right direction. There's no other appliance I know of that is so good at warming the smile and the soul, even as it dries your mittens and cooks your dinner!

Of course, cooking on a woodstove does take some measure of skill, but it's a pleasant, easily mastered task. I went from cooking with electricity to cooking on a wood range overnight, as it were. All you really need is a little time, patience and common sense.

Once you've got a good fire going with some solid hardwood chunks, you can get into some serious baking, like your grandmother used to do. Start with pies, breads and muffins at first, since these things bake well at higher temperatures. Afterwards, when the stove begins to cool, you can go into lightly-baked cakes and pastries, or casseroles you want to cook slowly as the hours go by. Of course, the long, slow fire is also ideal for re-warming pies on the cookstove's warming shelves, or simmering a delicious stew all day long!

Here is one of my favorite mid-winter woodstove meals for you to try. I hope it will provide the neophyte woodstove chef with the inspiration to conjure up many, many more.

Enjoy!

*Simple, Delicious Vegetable Casserole*

Saute:
¾ cup chopped onions
¼ cup chopped celery
1 cup chopped carrots
in some good safflower oil.
In a bowl, mix:
3 eggs
1 cup grated cheese (Muenster, Montery Jack or whatever)
¼ cup of wheat germ
½ tsp. salt (optional)
¼ tsp. each pepper, basil, thyme and other favorite herbs.

Combine veggies with egg and cheese mixture; stir until all is blended well. Pour into an oiled casserole or long baking pan. Pop in the woodstove oven for at least 1 hour, then test to see whether it's done. Let cool for 10 minutes before serving, then accompany the main dish with deep greens from the stovetop and a lush salad, if you like. And don't forget some hearty bread.

*Quick, Hearty Banana Bread*

Mix:
⅓ cup safflower oil
½ cup honey
3 large bananas, mashed
½ cup milk
2 eggs and/or ¼ cup soya flour mixed with one Tbsp. water.
Set aside.
Sift 3 cups of whole wheat flour with 1tsp. baking powder.

Add mixtures together and stir in ½ cup of sesame seeds. Pour all into oiled bread pans. Bake in the woodstove for 45–60 minutes.
Cool, slice and enjoy!

*For Desert: Baked Apples A La Woodstove*

Wash 8 medium-size apples, and core them only halfway through, so as not to puncture the skins at the bottom. Then mix:
¼ cup of dark honey
½ cup raisins
¼ cup of dark or light rum, *or* ¼ cup of sweet cider
1 tsp, cinnamon.

Let rest 10 minutes while you re-fuel the stove, then:

Remove the raisins from the liquid with a slotted spoon and put them into cored apples. Place the apples in a covered baking dish, then pour the liquid over all. Pop in a *hot* woodstove oven for 15 minutes, remove promptly when apples can be easily poked with a fork. Let the apples cool a minute, then serve with milk or ice cream.

*Butter Cookies*

Cream:
1 cup light honey
1 egg
¼ cup salt (optional)
2 tsp. vanilla extract.
Blend in 2⅔ cups whole wheat flour.

Mix well, then drop cookie dough from a spoon onto an oiled cookie sheet. Bake by the flames at 375°F. for 7 to 10 minutes.

—Sally May Harris

# (18) Completing the Cycle

I KNOW A WOMAN who cleaned the ashes out of her stove one Friday night, and carried them to the basement in a cardboard box before leaving the kids with the babysitter for an evening out with her husband. She returned to a blackened cellarhole where the house used to be. The children were saved, but all the family's worldly belongings were wiped out. "It's strange," she said later. "We have no ties to the past anymore. It's like a clean slate. We have a whole new beginning."

Unless you're looking for a new beginning in life, *never* collect or store wood ashes in paper bags or cardboard boxes —nothing, in fact, except a metal can. Still-glowing embers can ignite flammable products many hours after the fire has died out, turning your house into a raging inferno. When collecting wood ashes, as in every other aspect of woodburning, common sense safety measures are critical in making yours a pleasurable and profitable experience.

A good broad-bladed ash rake is handy for scraping ashes out of the stove if you don't have an ash pan, as is an ash shovel large enough to handle a good quantity of ash and still small enough to maneuver inside your stove's firebox. An old-fashioned coal bucket or other metal container is useful for carrying the ashes to their storage place, which should be a covered, galvanized steel barrel. Take this advice, use these devices, and you'll have no trouble with wood ashes burning down your house.

I know another woman, Peg Boyles by name, who is an outstanding organic gardener. "Wood ashes themselves are enough reason for the gardener to heat with wood," Peg says. "They are rich in potassium, trace elements, and provide a good source of phosphorous. They are valuable for neutralizing acid soils. To make full use of their value, though, wood ashes must be protected from the elements. Leaving them exposed to rain will leach out the potash."

Peg knows whereof she speaks. Ashes contain all the trace minerals that went into the wood in the first place, brought up to the surface from deep in the earth by the extensive root system of the tree. To waste these ashes after the wood is burned instead of putting them to good use is squandering an important natural resource.

Hardwood ash is considered superior to softwood ash by most gardeners, which is fortunate since the smart woodburner will be using hardwood almost exclusively. One cord of wood will yield some fifty to sixty pounds of ash, and this will lend a great deal of fertility to 500 square feet of lawn or garden space. To spread wood ashes in the garden, you can scatter them evenly by hand over the soil surface and till deeply,

*Building a Bonfire*

*As the sun begins to set, gather some dry cardboard boxes or other paper in a clear area free from overhanging branches.*

incorporating the ash into the soil well before planting, or you can use it as mulch around melon, squash and cucumber hills. A thin layer of ash can also be scattered over growing grass. Don't apply wood ashes to your potato patch, however, as the alkalinity will result in scabby potatoes that will not store well.

Wood ashes also make a good pest repellent. Mixed with an equal part of lime, or used alone, ashes can be sprinkled on plants to defend against many insect species. Surrounding a newly planted seedling with wood ashes can also discourage cutworms from felling young tomato plants.

The forest produces other products beneficial to the gardener as well. Leaves and needles, sawdust and brush that can be burned to ashes or converted to chips, are all useful in small scale food production. Not only do they help maintain fertility, they also condition the soil and help keep weeds down.

Fallen leaves are one of the gardener's richest and most abundant fertilizers and soil conditioners, Peg says. They collect dissolved nutrients the tree collects deep in the soil. By composting, tilling under, or mulching with leaves, bark, wood chips and sawdust, we can make the rich storehouses of plant nutrients available to our own crops. In many areas, you can collect copious quantities of leaves for the garden, and you won't even have to rake them yourself. Just let your neighbors know that you are willing to pick up their bagged leaves, and haul them off to your compost pile.

Using leaves in the garden greatly improves the condition of the soil, creating a deep, spongy growing medium which traps and distributes water, air and warmth. Increasing the organic matter in the soil has many advantages, and the use of fallen leaves is one of the best ways to accomplish this. Of course, large quantities of leaves should be composted or shredded before use; otherwise, the waxy surface of the leaves will pack together, producing an unpenetrable layer in the soil. If you want to till the leaves directly into the soil, spread them in a thin layer across the garden, and if you're using leaves from acidic species like oak or pine, add a handful of lime per bushel to neutralize them. The best bet is to use a variety of different kinds of leaves, though, to assure capturing a wide variety of nutrients.

Sawdust can also be a good soil additive, but even weath-

*As night dawns, build the brushpile on top of the wads of paper, with all branches running in the same direction. PACK DOWN THOROUGHLY.*

ered, well-aged wood dust breaks down very slowly in the soil, tying up large quantities of nitrogen that will not be available to your crops. The best way to use sawdust, if your lifestyle allows it, is to use it first as animal bedding, then till it in the garden. The large amounts of nitrogen present in the manure will offset the sawdust's storage characteristic, slowly releasing nitrogen to feed your garden plants for years to come.

Sawdust, wood chips and even bark are useful as surface mulch, and these products are now often sold for use in ornamental gardens. If you're cutting a lot of wood each year, you should have a pretty good home supply right in your backyard. If you've got plenty, you can even use these products as pathways, or if you can prevail upon a tree-chipping contractor to dump truckloads of chips in your backyard, as landfill for marshy areas.

When you've accumulated a large supply of treetops and other brush too small to cut into firewood, pile it all *in the same direction* compactly, then in an open area where there are no overhanging trees, touch off a brushfire. Burning brushpiles are fun to have a party around, and while the fire is adding a cheery glow to a nighttime celebration, it will also be producing fertilizer for the forest or garden, in effect completing nature's cycle.

In the end, most woodburners find themselves living a little closer to the land, and this means making the most of every available natural resource. While the forests provide abundant, renewable fuel supllies, the end product can enrich our gardens, and provide our own bodies with plenty of homegrown fuel.

*Touch off the pile, and soon you'll have a fire that will light up the evening. Next morning, rake together the ashes and scatter them on the garden.*

# (19) How to Select the Woodstove That's Right for You

As if you didn't have enough to think about while shopping for a woodburning heater—dealing with sales people, searching for the best buy, considering efficiency claims, and fiddling with baffle systems, dampers and thermostatic controls—most wood stoves come in *sizes* (small, medium and large; Kong, Son of Kong, Great Uncle of Kong; or whatever) and sooner or later you're going to have to decide which model is right for you.

If you're thinking of purchasing a prefabricated fireplace, Franklin unit or potbellied parlor stove to lend chilly winter evenings a cheery touch and a warm glow, you can pretty much forget about the heavy-duty brainstorming and concentrate on selecting a size that will fit into the space next to the chimney without setting grandpa's rocking chair on fire. But if you're considering a woodstove for your sole or primary source of heat, it's helpful to have some idea of the heating capacity you'll need.

No one yet has evolved a precise mathematical equation to determine exactly how many rooms of what size and configuration a given woodstove model will heat, and it's not likely anyone will do so within the foreseeable future. To make such a calculation, factors such as the number and type of windows in the house, the quality and quantity of insulation, and even the number, activities and lifestyles of family members would all have to be taken into consideration, in addition to the type, quality and moisture content of the firewood and the rate at which the wood is burned.

Woodburning is a high art that simply will not submit to such scientific calculations.

Now there are exact methods for calculating a house's heat containment properties and BTU requirements for establishing the desired temperature differential between outdoor and indoor climates on a certain blustery winter's day, and this information can be found in any standard heating engineer's manual. An engineer could, therefore, specify the exact size and type of stove best suited to your house (though the cost of such services might be greater than the price of the stove itself), but boning up on the calculus necessary to make the computation yourself might be enough to make you give up the idea of heating with wood entirely.

In the spirit of woodburning's artistic nature, therefore, we offer the following brief guide to arriving at a creative calculation of the woodstove size that's right for you.

The first step is to decide how much of your home you want to heat with wood and to calculate the number of cubic feet

*Building stoves; Warner Stove Co., Belmont, New Hampshire.*

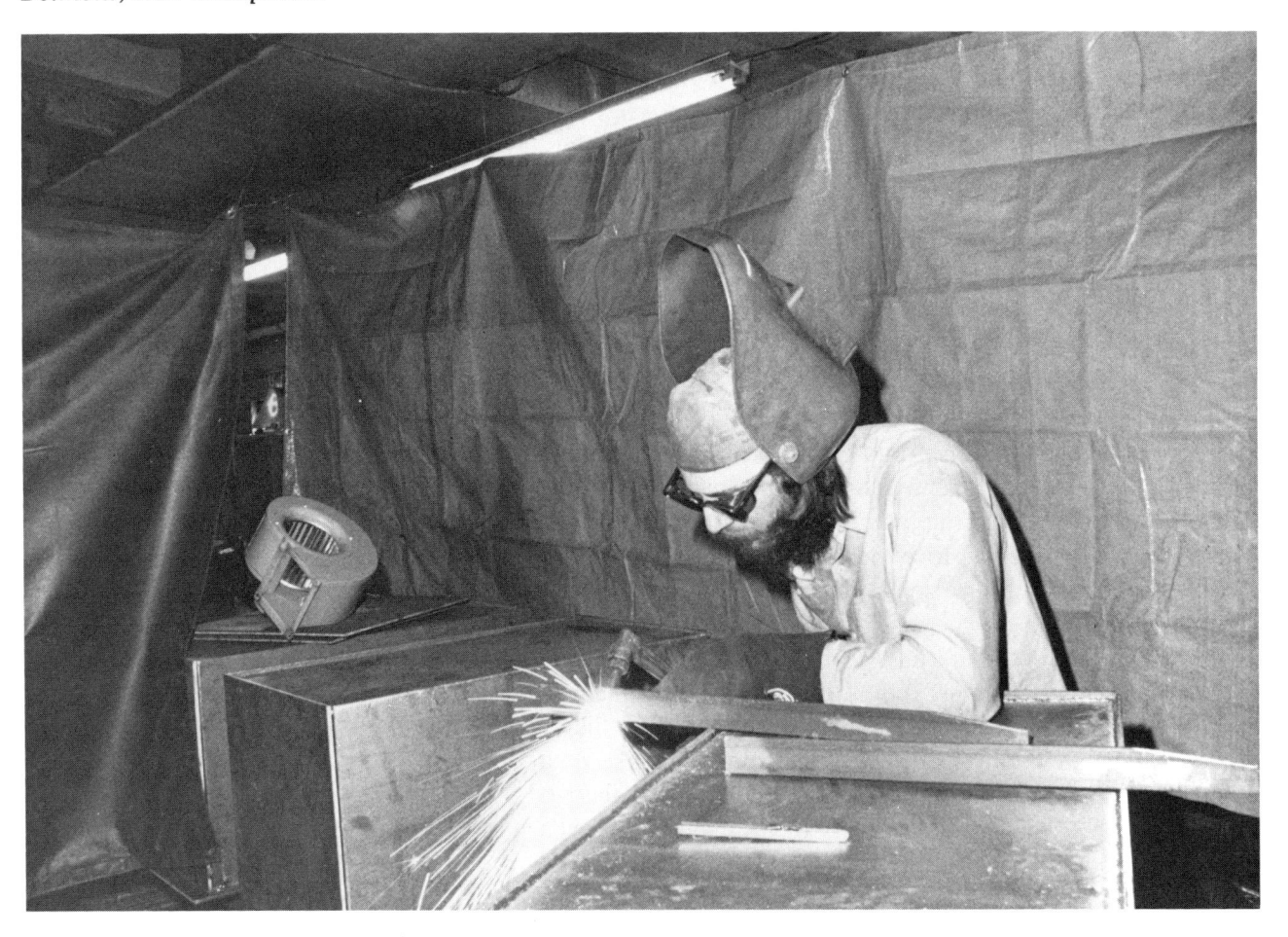

that area contains. If you plan to use wood heat as the sole or primary source of heat for the entire house, simply multiply its length by its width by the height of the ceilings. If you plan to heat both floors of a two-story house with wood, be sure to perform the same calculation for the second floor, and add the two results. If you're planning to heat only a portion of your house this way, measure the length, width and ceiling height for each room, stairway and hallway not closed off from the non-heated area, multiply the dimensions of each room in the same way, then add the results. *Example:* A single-story house 40 feet long by 25 feet wide with 8-foot ceilings: 40' × 25' × 8' = 8,000 cubic feet of space.

If you're fortunate enough to be considering a stove whose heating capacity is rated in cubic feet, your computation ends here and common sense takes over. If your house is totally uninsulated, it may be wise to purchase a stove with as much as twice the heating capacity this formula says you'll need, to be prepared for the worst of the winter weather. Since heat ratings for most stoves are based on optimum conditions, you should shave your estimation of your heating needs only if your home is extraordinarily well insulated and sealed tightly against air infiltration—and then only slightly.

When a stove manufacturer claims his product will warm "three to six rooms," he's not just avoiding the issue. The ability of a stove to heat a given area depends not only on the cubic feet of space inside, but also on the air flow and the contents of that space. One large room heats more easily than three smaller rooms of the same size; a room filled with furniture is slower to heat but holds the heat longer than an empty room; and snug, well-insulated rooms contain heat better than rooms with little or no insulation.

Again, common sense must rule the decision-making process, with a careful examination of your house's air flow pattern to lead the way. In a well-insulated house, all the average-size rooms with doorways opening to the room containing the woodstove should receive enough heat to keep their occupants comfortable, and second-story rooms directly above the woodstove area will receive their quota of warmth, especially if heating registers can be opened between the floors. Rooms separated from the stove by long hallways or more than two parallel walls, however, probably will not receive enough woodstove heat to provide the winter's warmth. To get some indication of your home's air flow, light a cigarette, incense stick or similar substance in the place where the stove will be located, and let it burn for about five minutes. If a member of your family with a keen sense of smell can detect the odor in

[144] HEATING WITH WOOD

*There are many alternatives for you in choosing the right stove.*

an outlying room, that room will probably be kept at a comfortable temperature by a well-fired woodstove.

Both the most scientific and the most troublesome indicator

of a stove's heating capacity is its BTU rating. It takes one BTU to raise the temperature of one cubic foot of living space by one degree Fahrenheit, on the average, so if you return to your 8,000-cubic-foot home some night and the inside temperature is 60°F. and you want to raise it to 70°F. in one hour, you'll need a stove that can generate 80,000 BTUs per hour, in addition to enough warmth to offset your rate of heat loss (which is dependent upon your insulation, windows, amount of exterior wall space and other factors).

*Calculation:* 8,000 cu.ft. × 1 BTU/cu.ft. × 10° temperature differential = 80,000 BTUs.

Although the absence of a precise heat loss rate makes this equation incomplete, it will provide some comprehension of the heating ability of a stove rated in BTUs per hour in relation to the cubic feet of living space inside your home. In addition, it is important to realize that the BTU output of a stove is usually based on its heat production while burning dry hardwood with the drafts fully open; that is, the *maximum* heating capacity of the stove. If you're firing your stove for a long, slow burn or using not quite dry wood for fuel, your hourly heat production may be three-quarters, one-half or one-fourth the advertised rating.

While you will have to turn to the engineering manual for a precise calculation of a stove's performance inside your home, these estimates can help you reach an informed decision about your personal heating needs. In the end, it's usually best to buy a stove slightly larger than you think you'll need; at the very least, you'll have a bigger firebox, which will mean longer burns between loadings and the capacity to really turn out the heat when that old north wind starts to blow.

## A FINAL NOTE

Before you declare your independence from the heating oil pipeline, and begin investing in a woodburner, tools and the other paraphernalia associated with wood heat, it's helpful to clearly and reasonably define your present and near future wood heating goals. With a firm idea of what you can logically hope to accomplish, given time, lifestyle, location and other constraints, you can best assure that your woodburning experience will be a pleasant profitable one.

Before investing in a woodstove, consider whether you do in fact want to use it as your primary source of heat, as an emergency heater to have on hand when the oil runs low, or to produce a cheery fire to take the chill out of a frigid midwinter evening.

If it's occasional wood heating you have in mind, you may

want to spare yourself the intricacies of a high-efficiency, airtight woodstove, and shop for a bargain in an attractive semi-antique parlor stove or an open hearth Franklin fireplace. Should you be shopping for an emergency woodheater, look to the inexpensive "barrel stoves" or even the do-it-yourself stove kits that are built around recycled 55-gallon oil drums. (Some of these stoves make up in efficiency what they lack in aesthetic appeal.) If you do intend to use wood as your primary source of heat, however, you'll be well-advised to purchase the best, most efficient woodburner you can find. Shop for heavy gauge cast iron or steel, tight-fitting doors and joints, a good baffle system and sturdy construction. Also watch for clean castings, smooth welds, and other signs of good workmanship. A good stove should last twenty-five years (look for the guarantee), and repay you many times the initial investment by pre-empting winter oil bills.

Lastly, you'll have to choose the type of *heat* you want from your stove. Radiant heaters actually warm the people and objects in the room, rather than the room air, and produce a more even, comfortable warmth. Convection heaters have a firebox surrounded by a steel cabinet. These units create warm air currents that help carry heat through the living space, but can also raise dust problems. Circulating heaters employ blowers and sometimes ducts to transfer heat to remote parts of the house.

# (20) How to Buy a Chain Saw

"IF I WERE TO move to an old-fashioned farm, everything quaint and handmade like a scene from Old Sturbridge Village, and could bring just one piece of modern machinery with me," Dartmouth College professor, writer, woodburner and part-time farmer Noel Perrin once observed, "I wouldn't hesitate a second. I'd bring my chain saw. It's noisy; it's dangerous; it pollutes the air—and I love it."

Loud, dirty and dangerous they may be, Professor, but nothing has done more to bring the age-old art of woodcutting to latter twentieth-century practicality than modern chain saws. Light years ahead of the earliest generations of chain saws that ripped and roared through the nation's timberlands, today's models are more subdued of voice, easier to handle and powerfully efficient.

Chain saw manufacturers offer a wide range of models and sizes. Before selecting one, it's important to know what saws are available, the jobs they are designed to do, and which features will be useful for the type of cutting you plan to do. "The Chain Saw Buyer's Guide" is intended as a handy introduction, a representative sampling of the choices available in the chain saw market. After considering the chart, visit local dealers, try out the demonstration models on hand, and take your time selecting the chain saw that's right for you. Chain saws are usually powered by gasoline engines ranging from a tiny 1.6 cubic inches to a monstrous six cubic inches in displacement. The cutting chains are mounted on bars that may be as short as ten inches in length for occasional, light-duty cutting, or as long as thirty-six inches. Most people find a sixteen-inch blade sufficient for cordwood cutting without being too awkward or unwieldy. The longer bars—needed for cutting only very large trees—can be dangerous in inexperienced hands.

Modern chain saws also come equipped with a wide range of options, ranging from handy automatic oilers with manual oiling overrides and safety chain brakes to luxuries such as heated handles, electric starters and automatic chain sharpening systems. While many of these extras may seem inviting, they can also complicate operation and maintenance of the basic machine. Experienced woodcutters have learned that

*A bewildering array of chain saws for sale.*

exotic frills can be more trouble than their convenience is worth, so careful, personal consideration of the options under inspection and actual cutting conditions at your dealer's woodpile is a good idea.

If you're in a position to invest in extras, one good buy to consider is a chain saw that has a slightly larger engine than is required for the bar length you choose. Not only will the extra power provide added protection against long-term wear and occasional strain on the saw, it will allow you to step up to a longer bar and chain, if needed—or add accessories to make your saw more versatile without buying a new chain saw later.

While testing a chain saw, don't forget to examine it for quiet running, low vibration and good balance of the saw's weight in your hands. When working in the woods, there's nothing that does more to bring on early fatigue than a whining, unmanageable machine.

If you are buying a saw for nothing more than some backyard tree pruning, or perhaps to cut up a few slender logs for the fireplace, a "mini-saw" will suit your purposes well. These saws are usually manufactured with engines of about two cubic inches displacement, and are equipped with guide bars ranging from ten to twelve or even fourteen inches in length. Mini-saws are light, easy to handle and inexpensive. Despite their diminutive size, many brands are also rugged and reliable. Before the newfound mass popularity of backyard chain sawing, these machines were manufactured for tree surgeons and orchardists who used them for treetop cutting when they had to climb to lofty heights with chain saws hanging from their belts to lop off limbs one-handed.

Many makers also offer electric mini-saws, and because they are less costly to manufacture, electric chain saws may sell for only about half the cost of equivalent gas-powered models. Electric saws must be plugged into a service outlet, so their use is limited by the length of your heavy-duty extension cord; however, if you are in the market for a handy, inexpensive tool for the sole purpose of whittling down the backyard woodpile, an electric saw might be a good investment. Not only are electric saws lighter, easier to start, cheaper and quieter than their gas-fueled counterparts, they produce no exhaust fumes and therefore can be used for woodcutting and construction projects in the woodshed or basement.

If you don't need to climb trees with your saw (a dangerous practice in the best of circumstances) and you'll be doing more than occasional cutting, a lightweight production chain saw might be the one for you. These saws are usually powered by engines in the 2.5 to 3 cubic-inch displacement range, and they are commonly outfitted with bars fourteen to sixteen inches in length. They hold more fuel than a mini-saw, and therefore run longer without refueling, but are quicker to start and easier to maneuver than the heavier production models.

Chain saws with displacements ranging from 3 to 4.9 cubic inches are known as medium-duty saws. These machines are best suited for regular logging and pulpwood cutting, and they will usually accept a wide range of brush-cutting, hedge, trimming, stonecutting and drilling attachments. Saws from this category are best suited to the year-round cutting done by professional wood suppliers, since they provide more power than is often required for cutting a single home's wood supply. These chain saws will certainly do the job around the homestead, but their size, weight and power may take a little getting used to.

Heavy-duty chain saws, the smallest of which have displacements of about five cubic inches and bars ranging from sixteen to thirty-six inches in length, are not for everyone. They are bigger heavier, stronger and louder than other production saws, and unless you plan to do contract logging in the White Mountain National Forest, convert your saw to an Alaskan sawmill for the production of lots of rough lumber, or cut giant trees every day, these formidable machines with their giant attachments may be more than you would care to handle.

If you're ready to buy a chain saw this year, first examine the following list of models, specifications and features to determine which machines might be right for you. Then, by all means, try before you buy. Ask questions. Talk with other woodcutters you know and the salesperson at your local chain saw dealer's showroom. After all else is carefully considered, make your final choice the saw that feels the best working in your hands. If all goes well, you'll be cutting with that saw for a long, long time.

# (21) Wood Splitters

THE TECHNOLOGY FOR splitting logs into firewood has come a long way since grandpa hung up his sledgehammer and retired his old iron wedges to a forgotten corner of the woodshed in favor of the standard splitting maul that's so widely used today. Properly applied to a row of two-foot cordwood chunks standing on end, the splitting hammer can be a swift and sure splitting "machine," but if you have a large collection of old roadside elm or knotty sugar maple, if you plan to produce more than half a dozen cords of firewood a year, or if you don't have the physical ability to split a lot of wood by hand, one of the more modern mechanical splitting devices may be for you.

There is a wide range of splitting equipment on the market today, from simple improvements on the basic splitting hammer to high-powered hydraulic splitters costing several thousand dollars. The simplest and least expensive device is the Chopper I, advertised as a radical improvement over the standard splitting tool. Shaped like a splitting maul, the Chopper I features a pair of hinged levers in its seven-pound head that deliver the force of the hammer's blow sideward to push the wood apart with relative ease. Its cost: less than $30.

Another splitting device that takes none of the labor out of hand splitting but removes some of the difficulty, danger and a bit of the sport is the Jiffy Woodsplitter. This gadget consists of a steel platform and a guidepost that holds its own wedge. A chunk of wood is placed on the platform and the wedge is positioned on top of it; then a blow from a sledgehammer finishes the job. The process is more time-consuming than freehand splitting, to be sure, but perhaps a bit safer for the woodburner who feels out of his or her element with tools in hand. The Jiffy Woodsplitter costs $89.95.

The lowest-cost power splitter on the market borrows its energy from the family car or any old clunker in the back field that can still spin its wheels. Produced under several brand names but known most commonly as The Stickler, this device consists of a single piece of steel shaped into a threaded cone with a flange on the end that bolts onto the car's rear hub after the tire and rim have been removed. (Some models come with a replaceable steel tip.) After the car has been jacked up so

*A hand-powered splitter.*

the left rear wheel is off the ground, you simply "feed" the unsplit logs to the mounted Stickler until the first threads bite into the wood. Within seconds, the screw-like threads will split the log in two, and the Stickler, still spinning, will be ready to take on another piece of wood. The cost of this splitting device is about $200, not including car and gasoline.

More sophisticated variations on this same theme are also on the market today. One, the Log Aug, can be operated without removing the auto tire. Its cost: about $389.

Standard hydraulic splitters powered by their own gasoline engines are the workhorses of the splitter field. These usually consist of a motor, a carriage for holding the log, and a large steel wedge that is thrust by hydraulic pressure through even the most stubborn chunk of wood and then quickly returned to its starting position. Hydraulic splitters may run from $700 for a five-horsepower unit that will deliver up to 22,000 pounds of force to a 21-inch (maximum) length log with a seven-second cycle time, to a mammoth 25-horsepower commercial model that can split 48-inch logs with four times the force in half the cycle time. These heavy-duty splitters often

*A hydraulic splitter.*

come equipped with electric starters and engines that sound like jet airplanes, and they can cost as much as $5,000.

Most tool rental centers have hydraulic splitters available on trailers designed to be towed behind cars. These gasoline-powered units, which commonly handle up to 26-inch long logs, rent for about $25 to $30 per day and can go through several cords of wood in that time.

The recent woodheating boom has spawned several improvements in the standard hydraulic splitter. One such device is the SS-500 Super Splitter, which features a five-way wedge that divides large diameter logs into five pieces in one thrust, instead of splitting them in half like ordinary splitters. The Super Splitter is designed primarily for professionals. Powered by a 25-horsepower engine, it takes only three and a half seconds to complete a splitting cycle and costs thousands of dollars.

Another variation of hydraulic splitter is the American brand vertical splitter. These devices work much like traditional splitting machines except that they thrust downward instead of horizontally to split the wood. The big advantage to the vertical splitter is that you don't have to lift large logs onto the carriage; instead, you can simply roll the log into position without ever hoisting if off the ground.

Firewood splitting will always involve a fair amount of hard work, whether the job is done by hand or with sophisticated hydraulic machines. Nevertheless, mechanical splitters can make an important contribution toward making firewood a practical source of large quantities of this inexpensive fuel, saving you time and money.

## FOR MORE INFORMATION

**The Stickler**
**Taos Equipment Co.**
Box 1565
Taos, N.M. 87571

**The Log Aug**
**El Fuego Corp.**
30 Lafayette Square
Vernon, Conn. 06066

**Bark-Buster** (a self-powered screw-type splitter)
**FW and Associates Inc.**
1855 Airport Rd.
Mansfield, Ohio 44903

**Futura Enterprises** (hydraulic splitters)
5069 Highway 45 South
West Bend, Wis. 53095

**The Super Splitter**
**LaFont Corp.**
1319 Town St.
Prentice, Wis. 54556

**American Woodsplitters Inc.** (vertical splitters)
Black River, N.Y. 13612

**Chopper I**
**Chopper Industries**
Box 87
Easton, Pa. 18042

**Cornell Manufacturing Co.**
Box 511, RD 2
Laceyville, Pa. 18623

**Lickity Splitter**
**Piqua Engineering**
Piqua, Ohio 45356

**Huss Sales and Service**
Toledo, Ohio 45356

**Knotty Wood Splitter Co.**
Hebron, Conn. 06248

**Didier Manufacturing Co.**
P.O. Box 163
Franksville, Wis. 53126

# THE BUYER'S GUIDES

# (1) The Wood Stove Buyer's Guide

FACTORS RANGING FROM safety and performance to construction, convenience, aesthetics and cost will influence the woodstove buyer's final decision, and no buyer's guide can substitute for a personal inspection of the various models of woodburners on the market today. Yet the following charts prove valuable by providing fingertip information for comparison of the ever-growing number of woodburners on the market and, it is hoped, by eventually helping you to choose a woodstove that will please when you buy it, and still prove satisfactory in the chill grip of winter. Every effort has been made to produce here the most comprehensive and complete woodbuyer's guide ever published.

## SOME TIPS TO REMEMBER BEFORE YOU BUY

In 1973, when the Original Energy Crisis caught America unaware of our abject dependence on fossil fuels, there were less than a dozen woodstove manufacturers left in the United States. Just a few short years after that Great Awakening, however, woodstove manufacturing is a multimillion-dollar business, and with the exception of some temporary summer oversupply problems during the past couple of years, it's been a constantly growing industry. Woodheater suppliers can now be found in cities and suburbs as well as rural areas. Surely, this is the latter twentieth-century renaissance of reliance on homegrown, renewable fuel.

By and large, this has been a boon to all parties concerned. Consumers have benefited by the larger choice of stove models available to choose from, and costs have been kept down somewhat due to economies of scale. The nation at large is also benefiting, through localized reductions in fossil fuel use and through improvements in our forest and fresh air environments. The future holds more benefits still.

Yet, as one major woodstove manufacturer noted: "When you have a booming market where anything will sell, you soon have a market where anything is for sale, from fine, durable, reliable heaters to junk. That's the situation right now."

This fellow wasn't talking about his own product, of course; no stovemaker would. Many woodstove manufacturers and dealers are honest, dependable businesspeople in the field for the long haul, so they work to maintain good reputations upon which their livelihoods depend. Still, you should approach each stove dealer with a skepticism you might ordinarily reserve for a used car salesman. Make more than one visit to several stove shops before you decide to buy, ask plenty of questions, and size up the salesperson's answers against what you already know—including some of the information contained in this book.

Since actual retail prices for woodstoves may vary widely, both above and below the suggested retail prices listed here, look around for value as well as selection. Try to buy stoves before the heating season arrives, or after it has temporarily passed, to avoid being stampeded into a purchase because of short supplies, and consider carefully features like construction materials, the size of a woodstove's loading door, whether or not it is thermostatically controlled or has a removable ash pan, and whether or not you'll need these features.

Inspect the stove carefully for good construction, efficient baffling system and workmanship, sealed joints and tightly closing draft regulators and doors. Kick the tires a bit, as it were, and bargain for the best deal you can get before you commit yourself to putting that sove in your living room.

*Caveat emptor* is the Golden Rule.

## WHAT MAKES A GOOD WOOD STOVE?

Considerations of safety having first been taken into account by the potential woodstove buyer, and personal evaluations of convenience, aesthetic, value and cost aside, the most frequent question asked about a woodheater's performance is its *efficiency*. After all, who doesn't want to get the most heat available out of every log he burns?

Unfortunately, the efficiency factor is exceedingly difficult to measure, much less compare among various stove models, because its computation involves variables ranging from combustion chamber designs to species and conditions of wood used as fuel. There are definite stove characteristics that dramatically affect heating efficiency, however. The most important are the following:

- *Airtightness:* Stoves that have tight-fitting joints and doors burn wood much more efficiently than stoves that allow unregulated air to seep into the firebox.
- *Complete combustion:* Among the stoves designed to be airtight, those that have effective baffle systems and/or smoke chambers will burn a higher percentage of volatile gases emitted from the wood and extract more heat from the same charge of wood than those that do not.
- *Surface area:* The larger the number of square inches of total surface area a stove presents to the air around it, the more heat the stove will radiate into the living space.
- *Conductivity and mass:* Certain stove materials, such as cast iron and steel, will readily conduct heat from the fire to the outside of the stove. Firebrick, glass and other insulating materials, while protecting the stove structure, are poor conductors of heat. Though they may be desirable for reasons of safety or aesthetics, they interfere with the stove's transfer of heat.

The mass of a stove's construction material also affects its performance. A stove made of sheet steel will be lighter than one constructed by heavy gauge cast iron. It will heat up faster, but will also lose heat faster after the fire goes out. Cast iron warms more slowly, radiates heat longer than steel, and is often considered more long-lived and durable.

| The Stoves | Construction | Guarantee | Mfr's Claims | The Models | Weight & Size (HxWxDepth) | Heating Capacity | Log Length | Price |
|---|---|---|---|---|---|---|---|---|
| All Nighter | Steel, firebrick lined, cast iron door, airtight type. | 30 days | UL approved for safety; patented air system provides 24-hr. burning. | Big Mo<br>Mid Mo<br>Little Mo<br>Tiny Mo | 524 lbs./31½x23½x41½<br>436 lbs./31½x21½x36<br>333 lbs./28x19½x32<br>267 lbs./26¾x17½x28½ | 7-10 rooms<br>5- 7 rooms<br>3- 5 rooms<br>1- 3 rooms | 28"<br>23"<br>18"<br>15" | $600<br>$530<br>$450<br>$370 |
| Alpiner | Firebrick lined steel and cast iron stepstove; airtight type. | 25 yrs. | Consistently offer more stove for the money. | Chamonix<br>Mont Blanc<br>Matterhorn | 325 lbs./28x16½x27<br>410/lbs.29¼x16½x34<br>485 lbs./29¼x18½x39 | 8,000 cu. ft.<br>11,000 cu. ft.<br>14,000 cu. ft. | 19"<br>27"<br>30" | $335<br>$385<br>$435 |
| Arctic | Cast iron, airtight type, baffled box heater. | 1 yr. | An extremely efficient, simply terrific stove. | 30 | 140 lbs./31x16x36 | None stated | 22" | $360 |
| Ardenne | Cast iron, firebrick lined box stove, airtight type. | 1 yr. | Solid, exceptional effectiveness with distinguished French design. | Ardenne | 259 lbs./28½x9½x25¾ | 10,000 cu. ft. | 24" | $450 |

# HEATING WITH WOOD

| The Stoves | Construction | Guarantee | Mfr's Claims | The Models | Weight & Size (HxWxDepth) | Heating Capacity | Log Length | Price |
|---|---|---|---|---|---|---|---|---|
| Ashley | Thermostatically controlled steel and cast iron circulating heaters. | None stated | Burns anything but rocks. | C-60<br>C-62<br>23-HF<br>23-HF | 242 lbs./36x35½x21¼<br>223 lbs./35x28¼x20¾<br>125 lbs./34x20x30<br>105 lbs./30x18x23½ | 4-5 rooms<br>3-4 rooms<br>4-5 rooms<br>1-3 rooms | 27"<br>19½"<br>24"<br>22¼" | $370<br>$325<br>$200<br>$185 |
| Atlanta | Thermostatically controlled steel circulating heater. | None stated | Efficiency, economy, durability. | Homesteader<br>24 WGE<br>Model 2502 | 240 lbs./33½x32¼x19¼<br>317 lbs./36½x35x20<br>147 lbs./35½x17½x22 | 4-5 rooms<br>4-5 rooms<br>3 rooms | 24"<br>24"<br>20" | $300<br>$525<br>$175 |
| Atlantic | Cast iron box stove, airtight type. | None stated | First original American cast iron, airtight box stove. | 224 | 180 lbs./24¼x20¾x35½ | 7,000 cu. ft. | 22" | $370 |
| Autocrat | Thermostatically controlled steel circulating heater. | None stated | 8-12 hrs. between fuelings. | Model FF76 | 245 lbs./34¼x32½x21¾ | 5 rooms | 25" | $425 |
| Birmingham | Thermostatically controlled steel circulating heater. | None stated | Efficient. One fueling lasts all night. | Knight 224<br>Knight 124<br>Majik 122-A | 317 lbs./36½x24x35<br>230 lbs./33½x32¼x18<br>156 lbs./35½x20x26 | 4-5 rooms<br>3-4 rooms<br>3-4 rooms | 24"<br>24"<br>22" | $450<br>$350<br>$225 |
| Bullard | Steel and firebrick, airtight type. | Lifetime | Energy efficient for fuel conservation. | The Eagle<br>The Hawk<br>The Falcon | 550 lbs./36x24x30<br>450 lbs./25x24x30<br>310 lbs./20½x18½x31 | 3,000 sq. ft.<br>8 rooms<br>3 rooms | 30"<br>22"<br>18" | $550<br>$525<br>$350 |
| Canadian Stepstove | Boilerplate steel, cast iron door, baffled. | Lifetime | Scandinavian design and efficiency at a domestic price. | Stepstove | 360 lbs./30x17x33 | 11,000 cu. ft. | 24" | $425 |
| Cawley/LeMay | All cast iron, airtight type box stoves with baffles and recessed cooking surfaces. | 25 yrs. | Uncompromising quality and efficiency. | 400<br>600 | 300 lbs./35½x18x36<br>385 lbs./35½x18x44 | 6,500 cu. ft.<br>10,000 cu. ft. | 16"<br>24" | $569<br>$699 |
| Chappee | Refractory lined steel with enamel finish and cast iron grate. | None stated | Cooktop surface; capable of efficiently heating small areas. | 8008 | 141 lbs./20¼x21x12½ | 4,240 cu. ft. | 14" | $325 |
| Comforter | All cast iron, airtight type. | 5 yrs. | Not only beautiful, but functions with highest efficiency and serves as excellent cooker. | The Parlor Stove | 270 lbs./26¾x24¼x21½ | 10,000 cu. ft. | 21" | $550 |
| Culvert Queen | Black steel culvert pipe with welded seams. | 3 yrs. | 55% efficiency. | The Culvert Queen | 140 lbs./31x33 dia. | 10,000 cu. ft. | 18" | $260 |
| DeDietrich | All cast iron, decoratively sculptured heater with oven and food warming area. | 1 yr. | True replica of 1684 French traditional model; efficient. | AL-77 | 240 lbs.39¾x28¾x18¼ | 19,840 BTUs/hr. | 20½" | $725 |
| Dover | Steel plate, lined with 3,000° refractory cement; heat exchanger included. | None stated | No other stove is as efficient. | Super Box | 225 lbs./36x22x44 | Still unknown | 28" | $525 |
| Dynamite | Boilerplate steel, airtight stove. Water heater available. | Unconditionally guarantted | Rugged. Burns long and slowly, uses less wood. | Dynamite<br>Greenwood Dynamite | 140 lbs./27x18x34<br>240 lbs./38x18x34 | 5,000 cu. ft.<br>12,000 cu. ft. | 24"<br>24" | $175<br>$250 |
| Elm | Cast iron, steel and firebrick baffled stoves with Pyrex viewing windows and cooking surface. | 1 yr. | Durable, steady fires for up to 14 hrs., and romantic view of the fire. | The Elm<br>The Short Elm | 275 lbs./26x23x33<br>240 lbs./26x23x27 | 7,000-9,000 cu. ft.<br>5,000-7,000 cu. ft. | 24"<br>18" | $425<br>$375 |
| Energy Harvester | All cast iron box stove; airtight type. | 5 yrs. | High-quality casting, baffled with dual controls for more efficient burning. | Chocorua | 220 lbs./26¾x18¼x34½ | 55,000 BTUs/hr. | 20" | $410 |
| Fatsco | Stainless steel body with optional cooking surface; can be used to barbecue. | None stated | Ideal for small spaces; efficient little cookers and heaters. | Tiny Tot<br>Pet<br>Chummy<br>Buddy | 15 lbs./14x10½ dia.<br>13 lbs./11¾x10½ dia.<br>25 lbs./15x10½ dia.<br>31 lbs./21½x10½ dia. | None stated<br>None stated<br>None stated<br>None stated | 8"<br>8"<br>8"<br>8" | $45<br>$40<br>$60<br>$110 |

THE WOODSTOVE BUYER'S GUIDE    [161]

[162] HEATING WITH WOOD

THE WOODSTOVE BUYER'S GUIDE [163]

| The Stoves | Construction | Guarantee | Mfr's Claims | The Models | Weight & Size (HxWxDepth) | Heating Capacity | Log Length | Price |
|---|---|---|---|---|---|---|---|---|
| Fisher | Welded steel stepstoves with cast iron doors; firebrick lined, two level cooking surface. | 25 yrs. | All day or night burning on one loading. | Baby Bear<br>Mama Bear<br>Papa Bear | 245 lbs./26¼x15½x28<br>345 lbs./30½x18x34<br>410 lbs./30½x20x39½ | 7,000 cu. ft.<br>10,500 cu. ft.<br>14,000 cu. ft. | 18″<br>24″<br>29″ | $350<br>$425<br>$475 |
| Fjord | All cast iron replica of Scandinavian airtight stoves. | None stated | Efficient, long-burning box stove. | Fjord | 231 lbs./30x14x31 | 8,000 BTUs/hr. | 24″ | $200 |
| Free Flow | Steel tubing body forms with baffles; airtight type. | 1 yr. | Far-reaching heat distribution and uniform temperatures . . . without electricity. | The Circulator<br>The Wonder<br>Furnace | 200 lbs./33x23x34<br>260 lbs./35x25x36<br>325 lbs./37x28x37 | 8,000 cu. ft.<br>12,000 cu. ft.<br>25,000 cu. ft. | 22½″<br>27½″<br>30″ | $480<br>$800<br>$720 |
| Frontier | Steel box stove. | As long as you own the stove. | Quality handcrafted, efficient; holds fire overnight. | Bx-24-6 | 250 lbs./23½x18x27 | 1,200 sq. ft. | 24″ | $300 |
| Hede | Fiberglass, insulated, enameled steel. | 1 yr. | Combines the best features of other wood stoves, with glass doors for viewing fire. | Hede | 240 lbs./42x31x20 | 8,000 cu. ft. | 24″ | $500 |
| Hinckley Shaker | Cast iron, steel baffles, airtight type. | None stated | 40% more heating surface area than other 2 ft. log burners. | Basic Shaker<br>Heat Exchanger | 225 lbs./25½x26x39½<br>265 lbs./32x26x39½ | 8,000-10,000 cu. ft.<br>8,000-10,000 cu. ft. | 26″<br>26″ | $425<br>$550 |
| Home Warmer | Thermostatically controlled steel and cast iron; airtight type. Extensive baffling. | Lifetime | Tested at 68.03% efficiency. | Home Warmer I<br>Home Warmer II | 320 lbs./33x18x30<br>270 lbs./28x16x24 | 60,000 BTUs/hr.<br>39,000 BTUs/hr. | 29″<br>23″ | $475<br>$410 |
| Huntsman | Welded steel, firebrick lined stepstove; airtight type. | Standard Sears home trial. | Efficient woodburner, burns combustion gases. | 42P84151N | 390 lbs./30x16x20 | None stated | 24″ | $325 |
| Independence | Welded steel. | 3 yrs. | 60% efficiency. | Independence<br>Independence Junior | 280 lbs./31x16x36<br>190 lbs./27x12x30 | 10,000 cu. ft.<br>6,000 cu. ft. | 28″<br>22″ | $400<br>$300 |
| Jotul | All cast iron, airtight type. Elaborate baffling. Optional enamel coatings. | 2 yrs. | Exceptionally long burning time; impressive fuel economy. | 118<br>602<br>606 | 231 lbs./30x14x31<br>117 lbs./25x13x21<br>175 lbs./41x12x22 | 8,980 BTUs/hr.<br>4,007 BTUs/hr.<br>5,565 BTUs/hr. | 24″<br>16″<br>16″ | $510<br>$280<br>$515 |
| Kachelofen | Custom designed ceramic tile stove. | 1 yr. | Efficient work of art, offers "healthiest type of heat available today." | Variety of styles | 1,000-2,000 lbs./ dimensions vary | 38,000 BTUs/hr. | 24″-30″ | $2,500 |
| Kickapoo | Boilerplate steel with cement-lined firebox and cast iron door and frame; airtight type. | 5 yrs. | Economical; model of quality design and craftsmanship. | BBRS<br>BBRA<br>BBRB<br>BBRC | 370 lbs./34¾x20¼x28<br>364 lbs./34x20x27<br>258 lbs./34x20x27<br>235 lbs./34x20x20 | 2,400 sq. ft.<br>6-8 rooms<br>1,000 sq. ft.<br>3-5 rooms | 24″<br>24″<br>24″<br>18″ | $515<br>$710<br>$510<br>$460 |
| | Boilerplate steel with firebrick lining; airtight type. | 5 yrs. | High quality, efficient heater. | Boxer | 330 lbs./34¼x21¼x26½ | 1,600 sq. ft. | 22″ | $460 |
| King | Thermostatically controlled steel circulating heater. | None stated | Up to 12 hrs. of constant, even heating. | 8801-B<br>9900-B | 240 lbs./32x33x21<br>267 lbs./34x33x24 | 3-4 rooms<br>4-5 rooms | 25½″<br>25½″ | $375<br>$400 |
| Koppe | Airtight, firebrick and cast iron ceramic tile stoves. Some models have windows for viewing fire. | None stated | Artistic masterpiece, well designed and well made. | KH 77<br>KH 56<br>KK 100/s<br>KK 150/s<br>KK 400 | 313 lbs./31x28x16<br>375 lbs./37x28x16<br>452 lbs./32x35x16<br>562 lbs./41x35x16<br>440 lbs./36x36x20 | 9,600 cu. ft.<br>10,700 cu. ft.<br>10,700 cu. ft.<br>14,400 cu. ft.<br>9,600 cu. ft. | 24″<br>24″<br>30″<br>30″<br>32″ | $735<br>$795<br>$980<br>$1,030<br>$1,000 |
| Lakewood | A-shaped steel and cast iron, firebrick lined stove; airtight type. | 5 yrs. | Full Scandinavian baffling system gives long burn. | Cottager<br>Workhorse | 250 lbs./25½x16½x29<br>500 lbs./37½x21x39½ | 1,000 sq. ft.<br>2,000 sq. ft. | 21″<br>32″ | $325<br>$550 |
| Lakewood Stepstove | Steel and cast iron, firebrick lined; airtight type. | 5 yrs. | Baffled for long burn, with two cooking surfaces. | The Stepstove | 350 lbs./30¼x17¾x33¼ | 1,500 sq. ft. | 26″ | $410 |

[164] HEATING WITH WOOD

| The Stoves | Construction | Guarantee | Mfr's Claims | The Models | Weight & Size (HxWxDepth) | Heating Capacity | Log Length | Price |
|---|---|---|---|---|---|---|---|---|
| Lange | All cast iron, elaborate baffling; airtight type. Optional enamel exterior. | None stated | Durable, safe, tightly made, producing fuel efficiency and long burning time. | 6303A<br>6303<br>6302A<br>6302K<br>6203BR<br>6204BR | 145 lbs./23½x16x25<br>220 lbs./37½x16x25<br>272 lbs./34x16x34<br>370 lbs./50½x16x34<br>213 lbs./41x13¼x20<br>250 lbs./41x13¼x25 | 3,000-5,000 cu. ft.<br>4,000-6,000 cu. ft.<br>7,000-9,000 cu. ft.<br>8,000-10,000 cu. ft.<br>4,000-6,000 cu. ft.<br>5,500-7,500 cu. ft. | 18"<br>18"<br>24"<br>24"<br>14"<br>18" | From $350<br>From $450<br>From $575<br>From $795<br>From $460<br>From $575 |
| Locke Stove | Steel, brick, cast iron grates. | None stated | Heavy-duty construction outclasses all similar stoves. | Warm Morning | 290 lbs./33x18x36 | 4-5 rooms | 24" | $550 |
| Montgomery Ward | Thermostatically controlled steel circulating heater. | Standard Ward's home trial. | Sensitive thermostat, circulates heat evenly. | 68A5718R<br>68A5722R | 212 lbs./33x32x21<br>285 lbs./34x32x21 | 3-4 rooms<br>4-5 rooms | 24"<br>25" | $265<br>$295 |
| Morso | All cast iron, airtight type. Optional enamel exterior. Some models with baffles. | None stated | Very efficient, economical. Elegant design, hand workmanship. Lasts a lifetime. | 1B<br>2B<br>1BO<br>2BO<br>6B | 254 lbs./34x14x30<br>124 lbs./28x13x27<br>353 lbs./51x14x30<br>164 lbs./40x13x27<br>146 lbs./24x14x24 | 6,000 cu. ft.<br>4,800 cu. ft.<br>9,000 cu. ft.<br>6,400 cu. ft.<br>4,300 cu. ft. | 22"<br>20"<br>22"<br>20"<br>18" | $475<br>$400<br>$775<br>$475<br>$535 |
| Nashua | Boilerplate steel, firebrick lined. Radiant plus circulatory heat design. | Lifetime guarantee | Alone in its field. Heats an entire house in five minutes flat. | Nashua 18<br>Nashua 24<br>Nashua 30 | 400 lbs./29¾x20¼x38¾<br>600 lbs./32¾x25x45½<br>1,045 lbs./40¾x31½x61½ | 8,000-12,000 cu. ft.<br>14,000-20,000 cu. ft.<br>21,000-35,000 cu. ft. | 18"<br>24"<br>30" | $595<br>$735<br>$1,395 |
| Norflame | All cast iron, double-walled and baffled, airtight type. | 5 yrs. | Produced in one of Europe's oldest foundries. | EX75 | 230 lbs./29½x14½x28 | 7,000 cu. ft. | 24" | $470 |
| Norman | All cast iron circulating heater, airtight type. | 1 yr. | Recirculation chamber makes this stove outperform others twice its size. | Norman 400 | 111 lbs./20x14x17 | 13,500 cu. ft. | 16" | $350 |
| Norwester | Thermostatically controlled steel and cast iron heater. | 1 yr. | Makes for a hotter and more efficient fire. | C-76 | 140 lbs./32¾x19x25 | None stated | 24" | $350 |
| Old Timer | Welded boilerplate steel with cast iron doors. Firebrick lined, baffled. | 5 yrs. | Extremely airtight, rugged and efficient. Cooking area. | Old Timer Stove | 455 lbs./35¾x19x37 | An Average home | 26" | $500 |
| Pillsbury | Steel and firebrick, cast iron door. Optional glass viewing door. | 25 yrs. | Very attractive, efficient baffled stove. Can fit into existing fireplace. | The Dove<br>The Falcon | 300 lbs./25x17¾x28¾<br>380 lbs./27x19¾x33¼ | 8,000 cu. ft.<br>11,000 cu. ft. | 20"<br>24" | $390<br>$440 |
| Prospector | Steel construction, firebrick lined, Vernier adjustments, high-capacity blower. | For life of original owner. | Air-tight construction; quality workmanship. | 1829 Prospector | 410 lbs./32½x19¾x43¼ | 15,000-28,000 cu. ft. | 24" | Not available |
| Quaker | Steel and cast iron, firebrick lined. | Lifetime | Special baffling system provides longer burn and excellent cooking efficiency. | Buck<br>Doe<br>Fawn | 510 lbs./34x17x35<br>480 lbs./34x15x32<br>380 lbs./34x14x21 | 17,000 cu. ft.<br>13,000 cu. ft.<br>9,000 cu. ft. | 28"<br>25"<br>15" | $515<br>$465<br>$415 |
| Ram | Welded steel stove, airtight. | 25 yrs. | Most efficient wood stove manufactured today. | Ram<br>Tile Stove | 250 lbs./30x14x36<br>300 lbs./30x14x36 | 15,000 cu. ft.<br>15,000 cu. ft. | 28½"<br>28½" | $460<br>$675 |
| Riteway | Thermostatically controlled steel and cast iron. Firebrick or aluminized steel linings. | 1 yr. | Heavy-duty construction for efficient burning and long life. | 2000<br>37 | 215 lbs./33x21x33<br>375 lbs./40x24x33 | 50,000 BTUs/hr.<br>75,000 BTUs/hr. | 24"<br>24" | $325<br>$460 |
| Scandia | All cast iron box stove; airtight type. | 5 yrs. | A rugged, economical, efficient Scandinavian stove. | 100<br>400 | 253 lbs./30x15x31<br>165 lbs./26x13x23 | 3-6 rooms<br>3-5 rooms | 26"<br>17" | $240<br>$155 |
| Sears Roebuck | Thermostatically controlled steel circulating heater. | Standard Sears home trial | Holds fire up to 12 hrs. without refueling. | 42G84123N<br>42G84065N | 147 lbs./35x20x27<br>235 lbs./33x32x19 | None stated<br>None stated | 24"<br>22" | $190<br>$250 |

# THE WOODSTOVE BUYER'S GUIDE [165]

| The Stoves | Construction | Guarantee | Mfr's Claims | The Models | Weight & Size (HxWxDepth) | Heating Capacity | Log Length | Price |
|---|---|---|---|---|---|---|---|---|
| Sevca | Recycled steel gas tanks. | None stated | Highest wood stove energy efficiency. | Double Barrel Stove<br>Baby Sevca | 255 lbs./36x16¾x36<br>175 lbs./36x16¾x24 | 50,000 BTUs/hr.<br>30,000 BTUs/hr. | 30"<br>18" | $350<br>$320 |
| Shenandoah | Thermostatically controlled steel and firebrick. R-76 circulating heater. | None stated | Economy, efficiency, durability, integrity. | R-76<br>R-77<br>R-65 | 260 lbs./36x24x35½<br>164 lbs./35x11x24 dia.<br>158 lbs./37x11x18 dia. | 4-6 rooms<br>1-3 rooms<br>1-3 rooms | 24"<br>21"<br>18" | $420<br>$315<br>$270 |
| Sierra | Steel and firebrick, cast iron door. | None stated | Very rugged, very efficient. | Sierra 300<br>Sierra 500 | 335 lbs./24x16x24<br>408 lbs./26½x18x32 | None stated<br>None stated | 24"<br>30" | $395<br>$445 |
| Styria | Firebrick lined, enameled steel with cast iron doors. | None stated | Superior stoves; handmade, efficient and safe. | Excelsior<br>Reliable<br>Imperial | 390 lbs./39x18x15<br>450 lbs./42½x18x15<br>640 lbs./48½x21¾x18½ | 1-2 rooms<br>2-3 rooms<br>3-5 rooms | 16"<br>16"<br>19¾" | $970<br>$820<br>$4,010 |
| Suburban | Thermostatically controlled steel circulating heater. | None stated | Economical, efficient. | Woodmaster | 225 lbs./32¼x32x19 | None stated | 23" | $275 |
| Sunshine | Steel and firebrick, airtight type. | 1 yr. | High efficiency. | 1 | 300 lbs./31x14x34 | 6,000-8,000 cu. ft. | 24" | $320 |
| Tempwood | Welded steel, airtight type, top-loading stove. | 15 yrs. | Efficiency, immediate heat and long burnability. | Tempwood II<br>Tempwood V | 200 lbs./28x28x18<br>100 lbs./24½x24x14 | 55,000 BTUs/hr.<br>35,000 BTUs/hr. | 18"<br>16" | $295<br>$345 |
| Thermo-Control | Steel and firebrick. Features built-in water heating system. Airtight type, may be connected to existing oil furnace or adapted for use in fireplace. | 10 yrs. | Thermostatically controlled downdraft burning produces high efficiency. | 100<br>200<br>400<br>500<br>TM2<br>TM4<br>TM5 | 200 lbs./27x18x33<br>210 lbs./27x18x33<br>375 lbs./27x24x34<br>425 lbs./33x24x40<br>210 lbs./29¼x22x36<br>325 lbs./29½x28x36<br>430 lbs./33x28x40 | 1-3 rooms<br>800-1,200 sq. ft.<br>1,000-1,600 sq. ft.<br>1,400-3,000 sq. ft.<br>1-3 rooms<br>2-4 rooms<br>2-6 rooms | 18"<br>18"<br>18"<br>24"<br>18"<br>18"<br>24" | $350<br>$375<br>$525<br>$660<br>$380<br>$525<br>$665 |
| Timberline | Steel step-type box stoves; firebrick lined, airtight type, baffled. | 5 yrs. | Rugged and efficient. | T-18<br>T-24<br>T-33 | 285 lbs./30x14x21<br>418 lbs./33x16x27<br>502 lbs./33x18x33 | 1,100 sq. ft.<br>1,600 sq. ft.<br>2,500 sq. ft. | 18"<br>24"<br>30" | $380<br>$450<br>$490 |
| Trolla | All cast iron, inside baffling and linings. Airtight type. | None stated | High fuel efficiency. Small stove does the work of larger conventional stoves. | 102<br>105<br>107 | 76 lbs./24½x11x17½<br>178 lbs./25½x13x21<br>253 lbs./28½x13x31 | 1,800-3,200 cu. ft.<br>3,000-5000 cu. ft.<br>4,000-7,000 cu. ft. | 14"<br>21"<br>28" | $280<br>$340<br>$490 |
| Ulefos | Airtight, baffled cast iron Scandinavian stoves. | 1 yr. | Precision-built, rugged, safe and efficient. | 868<br>864<br>865<br>172 | 115 lbs./24½x11x19<br>143 lbs./26x13x20½<br>253 lbs./31x14x33<br>374 lbs./68x13½x31 | 3,000 cu. ft.<br>4,000 cu. ft.<br>7,000-9,000 cu. ft.<br>9,000+ cu. ft. | 13½"<br>15½"<br>25½"<br>23" | $280<br>$310<br>$510<br>$900 |
| Vally Comfort | Steel and stainless steel. Airtight type. | 5 yrs. | Durable and highly efficient. | C-26<br>C-31 | 170 lbs./35x28x22<br>195 lbs./35x34x22 | 11,000 cu. ft.<br>15,000 cu. ft. | 18"<br>24" | $470<br>$485 |
| Vermont Downdrafter | Thermostatically controlled steel and firebrick heater with cast iron doors. Blower and hot water coils available. | 1 yr. | Only true downdraft stove on the market. Rated 60% efficiency. | DD1 | 500 lbs./32x26x34 | 60,000 BTUs/hr. | 24" | $715 |
| Warner | Boilerplate steel with cast iron door, baffled, airtight type. | Lifetime | Designed to be the most economical, efficient stove you can buy. | W118<br>W124<br>W130 | 285 lbs./25x15½x28¾<br>385 lbs./29½x18½x33½<br>510 lbs./29½x21½x42½ | 8,000 cu. ft.<br>12,000 cu. ft.<br>18,000 cu. ft. | 18"<br>24"<br>32" | $315<br>$390<br>$420 |
| Waverly | Steel circulating heater. | None stated | Provides circulating warm air. | 109 | 60 lbs./24x16x26 | 3,000 cu. ft. | 20" | $175 |
| Weso | Cast iron/ceramic tile stove. | 1 yr. | Efficient, beautiful, offers healthiest type of heat available today. | Weso | 500 lbs./32x34½x17½ | 6,400 cut. ft. min. | 18" | $1,015 |
| Wonderwood | Thermostatically controlled steel circulating heater. | None stated | Amazingly efficient. One fueling burns 10 hrs. | Model 2600 | 210 lbs./33x19x32½ | None stated | 24" | $305 |

[166] HEATING WITH WOOD

| The Stoves | Construction | Guarantee | Mfr's Claims | The Models | Weight & Size (HxWxDepth) | Heating Capacity | Log Length | Price |
|---|---|---|---|---|---|---|---|---|
| Wood King | Boilerplate steel with cast iron door and frame; airtight type. Top loading. | None stated | Extracts maximum heat from every fueling. | 2600 | 147 lbs./35x19x25 | None stated | 24" | $225 |
| Woodsman | Firebrick lined steel step-stove; nickle-plated cast iron doors. Airtight type. | 25 yrs. | A radiant heater and cookstove in one elegant unit. | A<br>B<br>C<br>D | 275 lbs./28x18x28<br>320 lbs./28x24x26<br>415 lbs./31x30x26<br>450 lbs./31x36x26 | 1,000 sq. ft.<br>1,500 sq. ft.<br>2,000 sq. ft.<br>3,000 sq. ft. | 22"<br>20"<br>21"<br>26" | $425<br>$525<br>$615<br>$650 |
| Yankee | Steel barrel-type stove with cast iron fittings. Stainless steel construction available.<br>Steel box stove; airtight type. | 120 days | Economical, efficient and reliable. | Model 20<br>Model 30<br><br>Model K | 75 lbs./36x18½x21<br>90 lbs./29½x18 dia.<br><br>225 lbs./28x20x36 | 7,000 cu. ft.<br>11,000 cu. ft.<br><br>15,000 cu. ft. | 18"<br>28"<br><br>30" | $150<br>$165<br><br>$495 |

## COOKSTOVES

The woodburning cookstove is a creation unto itself. Producer of the finest of breads, pies and roasts under the guidance of an expert master or mistress, the kitchen stove can also turn ornery, sullen and well-nigh impossible to light. Its primary purpose is to cook food, not warm people, and though the midwinter cook is sure to appreciate its cozy byproduct, the cookstove should be judged primarily on its ability to perform its intended chore.

The largest trade in kitchen woodstoves still seem to be the resale of refurbished nineteenth-century models—the most practical of antiques, except that these big cookers are not efficient woodburners. Increasingly, stove shops are stocking the smaller, frequently white-enameled twentieth-century improvements on these models, and new, traditional cookstoves

THE WOODSTOVE BUYER'S GUIDE    [167]

are now being cast by several American manufacturers today.

There are also some newcomers on the cookstove scene. Some Scandanavian manufacturers are producing modern, highly efficient kitchen stoves for the domestic market, and combination wood-electric and wood-gas cookstoves have appeared in response to the rising costs of fossil cooking fuels. A few European cookstoves are now also outfitted for heavy-duty, automatic domestic water heating.

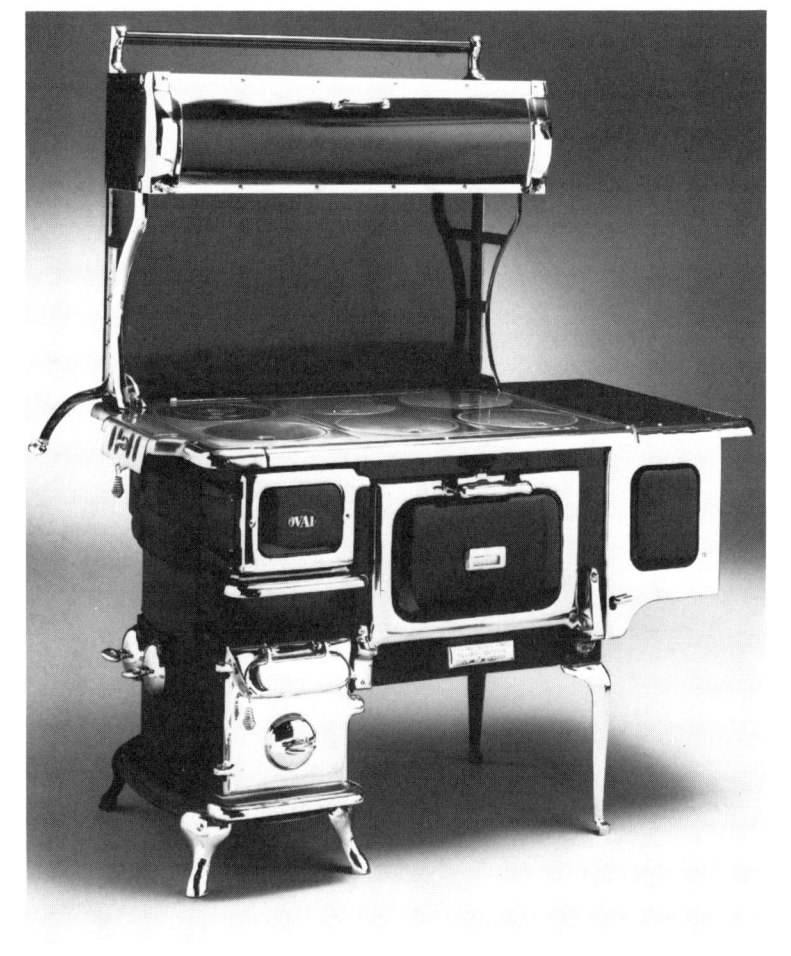

| Cookstoves | Construction | Guarantee | Mfr's Claims | The Models | Weight & Size (HxWxDepth) | Cooking Area | Log Length | Price |
|---|---|---|---|---|---|---|---|---|
| Atlanta | Solid cast iron. | None stated | Steady, even temperatures. Designed to last a lifetime. | 15-36 Range<br>8316 Cookstove | 254 lbs./29x35x21<br>155 lbs./28x30x21 | 6 covers, oven<br>4 covers, oven | 15"<br>15" | $550<br>$330 |
| Birmingham | Solid cast iron. | None stated | Long-lasting, even heat for cooking. | Red Mountain "T"<br>Bonanza | 300 lbs./29x35x21<br>165 lbs./28x30x21 | 6 covers, oven<br>4 covers, oven | 15"<br>15" | $435<br>$305 |
| De Dietrich | Cast iron; water jacket included. | None stated | Even heat oven, outer door stays cool; can be used to feed heat through domestic radiators. | 636 (other models available without water jacket) | 440 lbs./33½x31½x23¾ | Entire surface area | 12" | $1,025 |

| Cookstoves | Construction | Guarantee | Mfr's Claims | The Models | Weight & Size (HxWxDepth) | Cooking Area | Log Length | Price |
|---|---|---|---|---|---|---|---|---|
| Dynamite | Boilerplate steel, airtight stepstove. Water heater available. | Unconditionally guaranteed. | Can be loaded to burn through the night, or to hold a small cooking fire. | The Kitchen Cooker | 360 lbs./38x36x26 | Entire surface area | 24" | $375 |
| Findlay Oval | All cast iron, black porcelain finish with nickel plating with water tank. | Lifetime | The Cadillac, a top-quality stove. | The Oval | 400 lbs./58x48x24 | Entire surface | 16" | $1,710 |
| Jotul | All cast iron. | 2 yrs. | Airtight, efficient heating stove. | 404 (with oven) | 223 lbs./31½x25x17¾ | 2 large burners, small oven | 12" | $655 |
|  |  |  |  | 380 (without oven) | 131 lbs./27½x23½x16¼ | 1 large burner, 1 small burner | 18" | $375 |
| King | Steel and cast iron. | None stated | Dependable and economical. | Perfection | 202 lbs./25¼x22x15 | 4 covers, oven | 16" | $320 |
| Lange | Cast iron with brass rail surrounding cooking surface. | None stated | Large firebox. Will efficiently heat large areas. | 911W | 375 lbs./33¼x36¼x24 | Entire surface area | 10" | $760 |
| Monarch | Cast iron woodburner with 36" elec. or gas range and oven. | None stated | Unbreakable, quick heating, safe. | 6LEH (elec.) HG36HW (gas) | 325 lbs./47x36x26 325 lbs./47x36x26 | 4 burners, 2 covers, oven | 16" 16" | $875 $875 |
| Olympic | Steel with cast iron grates and cooking tops. | 1 yr. | Good design, excellent craftsmanship, painstaking attention to detail. | B-18-1 18-W 8-15 | 490 lbs./31½x35¼x26½ 320 lbs./32x32x22½ 245 lbs./30¾x34x23½ | 2 covers, end plate, oven | 16" 14" 15" | $1,110 $720 $420 |
| Pioneer | All cast iron; firebrick lined oven. Various size ovens, warmers and griddles available. | None stated | Classic 1911 reproductions with unequaled performance, durability and price. | Range | 325 lbs./30x30x28 | 4 covers, oven | 12" | $710 |
| Stanley | Steel. Airtight construction. | None stated | Fuel economy, excellent performance. | Mark I | 210 lbs./35x35x22 | 4 covers, oven | 12" | $820 |
| Styria | Steel, cast iron door, firebrick lined. Hot water tank and brick bread oven available. | None stated | Super efficient, holds heat overnight. | 130 119 106 | 1,100 lbs./34x51x40 800 lbs./34x47x36 611 lbs./34x42x34 | Entire surface area, with separate warming shelf and oven | 18" 18" 16½" | $2,375 $2,150 $1,945 |
| Tiba | Steel, firebrick, stainless steel or enamel finish. | None stated | Very efficient, stable cooking temperatures. | Tiba | 725 lbs./36x35x24 | 35x24 oven | 14" | $1,180 stainless $1,115 enamel |
| Tirolia | Cast iron and steel enamel finish. | 1 yr. | Completely insulated, completely airtight. Efficient space heating. | SD4 D6 D5N D7N D9N 9ZH | 245 lbs./28x26x20 310 lbs./31x33x21 355 lbs./33x29x23 420 lbs./33x35x23 540 lbs./33x43x23 551 lbs./34x35x23 | 18x20 oven 25x21 oven 21x23 oven 27x23 oven 35x23 oven 13x15x19 oven | 10" 12" 14" 14" 15" | $440 $650 $725 $820 $920 $1250 |
| Wamsler | Cast iron and steel; firebrick lined. | 30 days | Maximum efficiency and economy; airtight construction. | LSCK 90 (models with other options available) | 465 lbs./34x36x24 | Entire surface area | 14" | $670 |
| Wonderwood | Sheet steel body, cast iron top. | None stated | Rapid, even heating. | Ranch Stove | 105 lbs./29x27x20 | 4 covers, oven | 12" | $185 |

## FIREPLACE UNITS

These are the latest generation of devices designed to render the traditional fireplace more efficient. Unfettered by such improvements, the fireplace enriches the spirit and warms the heart, but the heat that it provides to the home is largely illusory. In fact, the functioning fireplace is often a net *extractor* of heat from the living space.

This situation can be remedied to varying degrees by the addition of circulating fireplaces or fireplace stoves. Some units fit into an existing fireplace, where the naked fire once burned, while others are designed to replace the fireplace in new home construction. Retrofitted fireplace units are *never* quite as efficient as the best, airtight woodstoves, but individual circumstances may make their purchase well worthwhile.

| Fireplace Units | Construction | Guarantee | Mfr's Claims | The Models | Weight & Size (HxWxDepth) | Heating Capacity | Log Length | Price |
|---|---|---|---|---|---|---|---|---|
| Charmaster | Steel with glass doors; water heating coil and blower available. | 1 yr. | Elegant, functional, economical, safe. | Charmaster II | 747 lbs./54x38x55 | Entire house | 30" | $1,960 |
| Dover | Boilerplate steel lined with refractory cement. | None stated | "We'll custom-build you a beautiful, heat-efficient fireplace stove." | Custom-built | Varies with installation 27x18x15 min. | Varies with size | Varies with size | Starts at $500 |
| El Fuego | Steel with tempered glass doors. | None stated | Firebox units for new or existing homes. Increase fireplace efficiency. | III (module) IV (pre-built) V (free-standing) | 3 sizes available to fit fireplace 325 lbs./48x44x24 250 lbs./48x44x28 | av. 42,000 BTUs/hr. | 18"-32" depending on model | $470 $770 $595 |

[170] HEATING WITH WOOD

| Fireplace Units | Construction | Guarantee | Mfr's Claims | The Models | Weight & Size (HxWxDepth) | Heating Capacity | Log Length | Price |
|---|---|---|---|---|---|---|---|---|
| Firemagic | Steel fireplace with cooking surface. | 5-yr. guarantee against warping. | Designed to produce the ultimate in heat and air circulation. | Independent | 300 lbs./32x28x26 | 1-3 rooms | 20" | $360 |
| Free Heat Machine | Steel heat exchanger with tempered glass doors. | 1 yr. | Five times more efficient than other systems. | 23-20<br>26-20<br>26-24<br>28-20<br>28-24 | 150 lbs./27½x24½x19¾<br>165 lbs./30½x24½x19¾<br>185 lbs./30½x24½x23¾<br>175 lbs./32½x24½x19¾<br>200 lbs./32½x24½x23¾ | 12,000 cu. ft.<br>12,000 cu. ft.<br>12,000 cu. ft.<br>12,000 cu. ft.<br>12,000 cu. ft. | 22"<br>22"<br>22"<br>22"<br>22" | $470<br>$470<br>$470<br>$520<br>$520 |
| Hearth Mate | Steel box stove with fireplace cover panel. | None stated | Converts the fireplace into an efficient home heater. | Better 'n Ben's | 150 lbs./24x18x24 | 10,000 cu. ft. | 18" | $325 |
| Heatilator | Steel circulating unit. | 90 days | A comforting supplemental heating system. | Mark 123C | 34½x46x21 (variable) | Depends on fire. | 30" | $330<br>$1,095 |
| Heatscreen Plus | Steel heat exchanger with tempered glass doors. | Standard Sears home trial | Reduced heat lost up chimney; channels heated air back into room. | 34P9702N2 | 109 lbs./24½-32¼x33¼-45x23 | None stated | 20" | $220 |
| Hydroplace | Double-walled water heating steel unit. Steel plate. | 25 yrs. | Can be used as sole source of home heat. | 40" Hydroplace | 450 lbs./56½x42x25 | 50,000 BTUs/hr. | Up to 40" | $940 |
| Leyden Hearth | Steel with tempered glass doors. | 2 yrs. | Distributes more heat over a wider area. Economical. | Leyden Hearth | 200 lbs./25x26x20 | 3 rooms | 22" | $520 |
| Majestic | Zero clearance prefabricated built-in fireplace system. | 1 yr. | 35% efficiency; UL approved. | ESF-II | 263 lbs./varies according to installation | None stated | 20" | $510 |
| Martin | Sheet steel with glass doors. | Limited warranty with 25-yr. protection plan. | Can be built anywhere you can dream of. | BW-28-45<br>BW-36-45A<br>BW-42-45A | 235 lbs./47x25x38<br>287 lbs./49½x25x45<br>391 lbs./54x25x51 | 42,600 BTUs/hr. | 28"<br>36"<br>42" | $295<br>$345<br>$480 |
| Thermo-Control | Thermostatically controlled steel box stove with fireplace cover panel. | None stated | Airtight, efficient heating system. | Model 300 | 240 lbs./28x18x24 | 3-4 rooms | 22" | $425 |
| Thriftchanger | Fireplace is refractory brick, with tempered glass folding doors. Steel heat exchangers. | None stated | Highly efficient passive heating system. | Thriftchanger heat recovery fireplace | Variable/adaptable | Entire house. | 24"-32" | Less than $2,000<br>$2,025 |
| Timberline | Airtight steel and firebrick stove that fits standard fireplace. Blower optional. | 5 yrs. | Rugged and efficient. Holds fire overnight. | TFI | 32x23x22 | 1,800 sq. ft. | 20" | $625 |
| Yankee | Steel with cast iron doors; chimney baffle not included. | 120 days | Designed exclusively to fit existing fireplaces. | F | 180 lbs./22x30½x22 | None stated | 24" | $315 |

## WOOD BOILERS AND HOT AIR FURNACES

It may well be that efficient woodburning boilers and furnaces will be the home heating appliances of the future. Heating a home with woodstoves allows for greater flexibility in the choice of which areas are to be closed off from the heat flow and abandoned for the winter, but basement woodburners provide the comfort of central heating throughout the house. Heating an entire home with a boiler or hot air furnace requires a substantial initial investment for the heating plant and necessary ductwork, however, and also consumes large volumes of cordwood each winter.

In addition to the advantages of comfort, these woodburners are convenient. A central wood heating system means there is only one heater to feed—usually no more than two or three times in twenty-four hours—and all the bark chips, wood splinters, ashes and associated mess that goes along with heating with wood can be confined to the basement. Wood-fired furnaces can easily be adapted to provide domestic hot water too.

| Boilers and Furnaces | Construction | Guarantee | Mfr's Claims | The Models | Weight & Size (HxWxDepth) | Heating Capacity | Log Length | Price |
|---|---|---|---|---|---|---|---|---|
| Bellway | Welded steel hot air furnace. Hot water coils, automatic non-electric humidifiers and non-electric thermostats available. | None stated | Ten hours of heat from one fueling. | Four sizes available | Weight varies 48-60x30-46x44-72 | 10 rooms | 24"-48" | $625 $3,590 |

[172] HEATING WITH WOOD

| Boilers and Furnaces | Construction | Guarantee | Mfr's Claims | The Models | Weight & Size (HxWxDepth) | Heating Capacity | Log Length | Price |
|---|---|---|---|---|---|---|---|---|
| Charmaster | Thermostatically controlled, welded steel hot air furnace. | 1 yr. | Converts wood to charcoal, burning gases produced at high temperatures. | Charmaster Furnace | 646 lbs./54x28x55 | Entire house | 30" | $1,295 |
| Combo | Steel, wood-only and oil-wood boilers and hot air furnaces with optional domestic water heating. Lined with serra felt instead of firebrick. | 1-20 yrs. on various parts | Efficient. Special quality and safety features. | WO 12 23 1<br>WO 12 23 3<br>W 12 22 6<br>WO B 22 11<br>WO B 22 13 | 525 lbs./53x28½x54<br>585 lbs./53x28½x57<br>620 lbs./52x28½x67<br>820 lbs./49½x28½x30½<br>870 lbs./49½x28½x35½ | 95,000 BTUs/hr.<br>126,000 BTUs/hr.<br>140,000 BTUs/hr.<br>126,000 BTUs/hr.<br>156,000 BTUs/hr. | 22"<br>24"<br>28"<br>22"<br>28" | $1,270<br>$1,320<br>$1,090<br>$1,610<br>$1,695 |
| Controlled Combustion Systems | Welded steel water boiler with baffle system. | 5 yrs. | Patented combustion system burns up to 48 hrs. on one loading. | A<br>B | 1,250 lbs./54x30x60<br>950 lbs./54x30x30 | Entire house | 48"<br>24" | $2,970<br>$2,470 |
| Daniels | Welded steel and cast iron, insulated wood and wood-oil furnaces. | None stated | The original chunk wood furnace. Built right to work right. | R30W<br>R30WO | 450 lbs./41x36x41<br>450 lbs./41x36x41 | 10,000 cu. ft.<br>10,000 cu. ft. | 28"<br>28" | $1,250 (wood)<br>$2,000 (wood-oil) |
| Dover | Boilerplate steel lined with refractory cement. Blower and thermostat available. | None stated | Operates independently or attached to existing oil or gas furnace. | Super Box | 600 lbs./55x41¾x34¼ | Entire house | 24" | $745 |
| Dynamite | Boilerplate steel, airtight wood furnace. Water heater available. | Unconditionally guaranteed. | Rugged, reliable, recommended by owners to others. | Greenwood Furnace | 300 lbs./38x18x34 | Up to 30,000 cu. ft. | 20" | $240 |
| Hoval Boiler | Steel-plated boiler, Swiss manufactured. | 5 yrs. (1 yr. on controls) | Efficiency: 87% with oil and gas; 74% with wood Incl. domestic hot water. | Six models | Average: 1,400 lbs./70x33x28 with hot water tank | 100,000 BTUs | 14"-21" | $1,625<br>$2,725 |
| Hunter | Stainless steel firebox, steel plate. | 1 yr. | Economical multi-fuel heating system. | HWO-100 | 790 lbs./53x48x56 | 140,000 BTUs/hr. | 28" | $1,520 |
| Independence | Welded sheet steel hot air furnace with manual draft; automatic draft optional. | 3 yrs. | 55% efficiency. | A<br>B<br>C<br>D<br>E<br>F | 300 lbs./30x22 dia.<br>325 lbs./30x24 dia.<br>350 lbs./30x26 dia.<br>325 lbs./36x22 dia.<br>350 lbs./36x24 dia.<br>375 lbs./36x26 dia. | All models 90,000-150,000 BTUs/hr. | 24"<br>24"<br>24"<br>34"<br>34"<br>34" | $415<br>$445<br>$475<br>$445<br>$475<br>$510 |
| Kerr | Steel boilerplate boiler or furnace. | 5 yrs. | Economical & dependable. | Titan | 675 lbs./43½x26½x35½ | Entire house | 24" | $1250 |
| Kickapoo | Steel hot air furnace. Blower optional. | 5 yrs. | Effective alone or combined with gas, oil, electric or solar heat. | BBR-D | 334 lbs./35¼x26x30 | 2,400 sq. ft. | 24" | $585 |
| Logwood | Welded steel hot air furnace. Optional oil burner. | 5 yrs. | Heaviest gauge burner on the market. | 36<br>24 | 1,200 lbs./61x32x60<br>1,000 lbs./61x32x48 | 200,000 BTUs<br>140,000 BTUs | 36"<br>24" | $1,845<br>$1,515 |
| Longwood | Reinforced steel hot air furnace. | 10 yrs. pro-rated | Most economical furnace on the market. | Wood-Oil-or Gas Combination Furnace | 700 lbs./40x24x66 | 150,000 BTUs | 5' | $1,425<br>$1,525 |
| Lynndale | Steel and firebrick; cast iron grates. | None stated | Central heating system, maximum heating utilization. | 810 | 1,200 lbs./51x34x68 | 125,000 BTUs/hr. | 36" | $2,025 |
| Monarch | Steel and cast iron, thermostatically controled, firebrick lined hot air wood furnace. Blower optional. | 1 yr. | Connects to existing gas or oil furnace. Maximum utilization of fuel load. | AF324 | 300 lbs./42x22x32 | 75,000 BTUs/hr. | 24" | $665 |

THE WOODSTOVE BUYER'S GUIDE

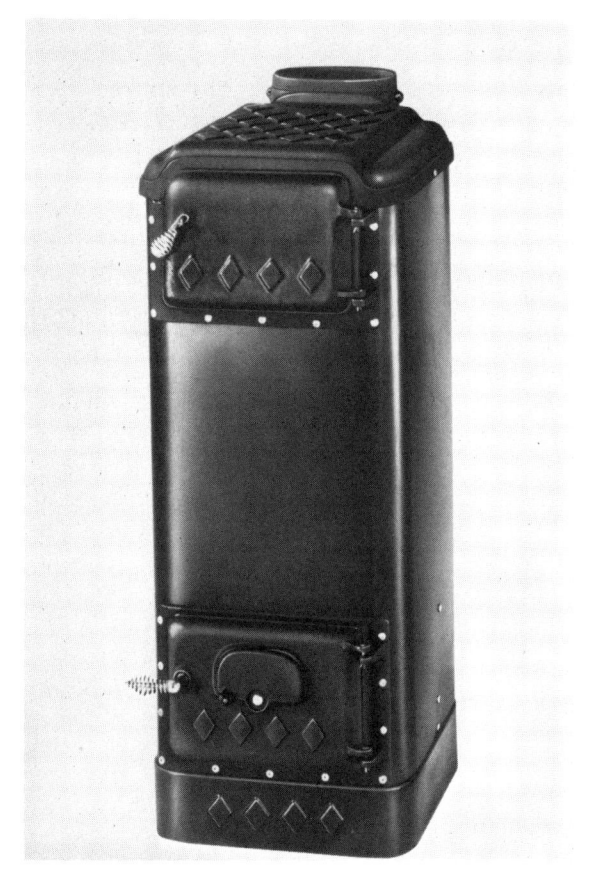

| Boilers and Furnaces | Construction | Guarantee | Mfr's Claims | The Models | Weight & Size (HxWxDepth) | Heating Capacity | Log Length | Price |
|---|---|---|---|---|---|---|---|---|
| Newmac | Welded steel hot air furnace. | 10 yrs. | Top woodburner in the USA. | Wood-Oil Combination | 1,100 lbs./50x48x54 | 167,000 BTUs | 18" | $1,620 |
| Northeaster | Cold rolled steel. | None stated | Super-efficient hot air furnace. | 224B<br>101-B | 550 lbs./48x36x35<br>460 lbs./48x36x29 | 125,000 BTUs/hr.<br>100,000 BTUs/hr. | 24"<br>18" | $1,270<br>$1,120 |
| Oneida Royal | Steel and firebrick, thermostatically controlled wood and wood-coal-oil-gas boilers and hot air furnaces. | 10 yrs. | Provides economical, trouble-free home heating. | WOB112 (boiler)<br>Woodcraft 120 (attaches to existing furnace)<br>"All Fuel" AF 160<br>"Two in One" WGO 160 | 1,483 lbs./47¼x21x42½<br>475 lbs./54x24x35<br><br>1,200 lbs./65½x30¾x67<br>1,360 lbs./59½x30¾x67 | Entire house<br>Entire house<br><br>Entire house<br>Entire house | 20"<br>24"<br><br>22"<br>22" | $4,225<br>$1,070<br><br>$2,400<br>$2,620 |
| Passat | Welded steel combination hot air furnaces and boilers. Optional blowers. | 10 yrs. | Will burn large chunks of unsplit wood, also paper, cardboard and anything else. | HO20 (boiler)<br>HOL20 (hot air) | 442 lbs./35x29x48<br>600 lbs./55x29x72 | 72,000 BTUs/hr.<br>88,000 BTUs/hr. | 36"<br>36" | $1,175<br>$1,615 |
| Ram | Welded steel hot air furnace and boiler. | 25 yrs. | Combines with existing oil furnace or boiler. | Ram (furnace)<br>Ram (boiler) | 350 lbs./48x27x42<br>350 lbs./30x16x38 | Entire house<br>Entire house | 28"<br>28" | $620<br>$740 |
| Riteway | Steel, firebrick lined wood-oil boilers and hot air furnaces. | 1 yr. | Well constructed for long service and efficiency. | 4 hot air furnaces<br><br>4 water boilers | Vary according to model<br>Vary according to model | 125,000 BTUs/hr.<br><br>350,000 BTUs/hr. | Varies<br><br>Varies | $1,420<br>$2,720<br>$2,420<br>$3,820 |
| Spaulding | Welded steel hot air furnace. | None stated | 75% of the energy out of nearly anything. | The Spaulding | 350 lbs./72x24x38 | 100,000 BTUs | 200 lbs. of anything | $2,335 |
| Tarm | Steel oil-wood boilers with optional domestic water heating.<br><br>Steel wet-base burner. | 5 yrs. | Energy-conserving, economical, strong. | OT 35<br>OT 50<br>OT 70<br>OT 28<br><br>MBS-55X | 1,100 lbs./38x39x27<br>1,500 lbs./37x46x27<br>1,850 lbs./37x46x36<br>900 lbs./38x39½x19<br><br>1,100 lbs./44x15½x41 | 112,000 BTUs<br>140,000 BTUs<br>200,000 BTUs<br>71,430 BTUs (wood)<br><br>140,000 BTUs | 18"<br>18"<br>28"<br>10"<br><br>27½" | $2,025-<br>$3,050<br>(varies with installation)<br>$1,575 |
| Tasso | Cast iron wood boiler. | None stated | Well insulated; lasts a lifetime. | Series with 8-11 sections available | 787-1,083 lbs. 40x18x12-27 | 126,000-185,000 BTUs/hr. | 20"-30" | $1,270<br>$1,575 |
| Thermo Pride | Welded steel, firebrick lined hot air furnace. | 10 yrs. | Burns wood or coal with thermostatically controlled efficiency. | W/C 20<br>W/C 27 | 635 lbs./43¼x25x50¼<br>850 lbs./46¾x27x58½ | 90,000 BTUs/hr.<br>130,000 BTUs/hr. | 20"<br>27" | $1,595<br>$1,745 |
| Valley Comfort | Welded steel hot air furnaces. | None stated | Better heat at lower cost. | RB-3D<br>RB-4D | 485 lbs./46x25x36<br>580 lbs./46x25x48 | 90,000 BTUs<br>120,000 BTUs | 33"<br>45" | $910<br>$965 |
| Volcano | Thermostatically controlled, welded steel. | 5 yrs. | True secondary combustion visible through an inspection port. | Volcano II (hot air)<br>Volcano III (boiler) | 585 lbs./42x21x29<br><br>585 lbs./38x21x28 | 120,000 BTUs/hr.<br><br>120,000 BTUs/hr. | 24"<br><br>24" | $705<br><br>$785 |
| Yankee | Firebrick lined steel barrel; cast iron door. | 120 days | Connects to existing hot water or forced hot air systems. | B (water system)<br>R (hot air system) | 642 lbs./36½x30x39½<br>496 lbs./36½x30x39½ | 125,000 BTUs/hr.<br>125,000 BTUs/hr. | 36"<br><br>36" | $1,010<br><br>$620 |
| Yukon | Steel cast iron and firebrick hot air furnace. | None stated | Automatic firing oil-wood furnace. | LW085<br>LW0100<br>LW0112 | 905 lbs./50x20x30<br>905 lbs./50x24x30<br>905 lbs./50x30x30 | 106,000 BTUs/hr.<br>125,000 BTUs/hr.<br>140,000 BTUs/hr. | 22"<br>22"<br>22" | $1,325<br>$1,720 |

## WOODSTOVE KITS

One of the newest developments to emerge from the wood stove market is the do-it-yourself stove kit. These kits range from an enameled, Scandinavian-type box stove to barrel heaters made from recycled oil drums. Generally, these units can all result in efficient, low cost woodburning if care is taken during construction.

Besides the satisfaction of building your own woodstove, do-it-yourself stove kits have the advantage of lower retail prices than their ready-made counterparts.

| The Kits | Construction | Guarantee | Mfr's Claims | The Models | Contents | Heating Capacity | Log Length | Price |
|---|---|---|---|---|---|---|---|---|
| Country Craftsmen | Cast iron door assembly and fuel; steel legs. | Lifetime | Ugly yet efficient; the lowest priced cast iron kit on the market today. | Model 15/55 | Everything needed but drum. | 1-3 rooms | 28"-34" depending on size of drum used. | $42 |
| Fatsco | Steel barrel and grease drum assemblies. | None stated | Neat, serviceable, long-life stove that assembles easily. | Woodsman | Everything needed but drum. | None stated | Depends on size of drum used. | $70 |
| Fisher's | Steel barrel and grease drum assembly with Pyrex glass door for viewing fire. | None stated | Functional, eye-appealing, easy to assemble, inexpensive. | Kits to fit 15-, 30- or 55-gallon drum | Glass door, legs, finishing materials. | None stated | Depends on size of drum used. | $42 |
| Reginald | All cast iron box stove; airtight type. | None stated | Elegant, durable, efficient. | Reginald 101 | Entire stove. | 4,700 cu. ft. | 16" | $280 |
| Washington | Steel barrel and grease drum assemblies. | 1 yr. | You can have a sturdy, rustic stove that will last several seasons. | Oil drum | Doors, legs and flue collar. | None stated | Depends on size of drum used. | $95 |
| Yankee | Steel and firebrick box stove; airtight type. Cast iron and steel barrel assembly. | 120 days | Rugged and dependable; perfect for do-it-yourselfer. | Model K<br>Model 20<br>Model 30 | Entire stove.<br>Entire stove.<br>Entire stove. | 15,000 cu. ft.<br>7,000 cu. ft.<br>11,000 cu. ft. | 30"<br>18"<br>28" | $260<br>$95<br>$110 |

## COMBINATION STOVES

The Franklin, or "combination stove" as more efficient units of this genre are known today, bears little resemblance to the "Pennsylvania Fire-Place" first invented by the great American statesman in 1740. Benjamin Franklin set out to improve upon the efficiency of the traditional fireplace, but being culturally English, insisted on being able to view the fire burning inside his creation.

Today's Franklin stoves achieve both of their namesake's aims, with varying degrees of success. Open, the combination stoves provide the cheery warmth of an unfettered fire. With doors closed, some become efficient woodheaters, although only a few models approach the efficiency of the standard woodburning stove. Traditional Franklin stoves are a compromise between raw efficiency and pure aesthetics, and function as well as a marriage between these two virtues might be expected to perform.

| Combination Stoves | Construction | Guarantee | Mfr's Claims | The Models | Weight & Size (HxWxDepth) | Heating Capacity | Log Length | Price |
|---|---|---|---|---|---|---|---|---|
| AFS | Thermostatically controlled steel and firebrick; cast iron doors. | 25 yrs. | Baffle design allows more heat from less fuel. | THE AFS Fireplace | 400 lbs./28x26x36 | 2,400 sq. ft. | 21" | $535 |
| Atlanta | Solid cast iron. | None stated | Lifetime use for heating or cooking. | 32<br>26<br>22 | 400 lbs./32¾x44x30<br>310 lbs./32x28x35<br>248 lbs./30x34x23 | 3 rooms<br>2-3 rooms<br>2 rooms | 27"<br>23"<br>18" | $545<br>$405<br>$530 |
| Atlantic | Cast iron, airtight. | None stated | Precision-fitted castings. Horizontal baffling system combined with time-tested draft control means efficient overnight burning. | 228 | 280 lbs./31x28¾x21¼ | 12,000-14,000 cu. ft. | 24" | $535 |
| Autocrat | Steel cabinet, cast iron linings. Thermostatically controlled. | None stated | Welded seams, seals tightly. | Americana | 400 lbs./36¼x42¾x29 | 5-6 rooms | 24" | $770 |
| Cherokee | Steel double walls with baffle system. | Lifetime | 80% efficient—60% with doors open. | Cherokee Chief<br>Cherokee Princess | 360 lbs./25x33¾x17<br>345 lbs./22x29½x16 | 3,000 sq. ft.<br>2,700 sq. ft. | None stated | $690<br>$675 |
| Comforter | All cast iron, airtight type. | 5 yrs. | Air preheat system, baffle with Venturi for secondary air combustion means efficient burning. | Fireplace | 270 lbs./26¾x24¼x21½ | 10,000 cu. ft. | 21" | $565 |
| Dover | Welded steel and cast iron. | None stated | Solidly built. Heating pipes increases efficiency. | The Dover | 325 lbs./32½x33¼x32¼ | 6-7 rooms | 24" | $370 |
| Dynamite | Boilerplate steel, airtight stove. Water heater available. | Unconditionally guaranteed. | Enjoys an extraordinary reputation among people who know wood stoves. | The Fireplace Stove (small)<br>The Fireplace Stove (larger) | 140 lbs./27x18x34<br>240 lbs./31x18x39 | 8,000 cu. ft.<br>15,000 cu. ft. | 24"<br>24" | $240<br>$340 |
| Efel Kamina | Porcelained steel and cast iron Pyrex glass door, airtight type. Six colors available. | None stated | Long-burning heater, includes cooking top and barbecue pit. | Efel | 199 lbs./32¼x28x15 | 16,000 cu. ft. | 18" | $545 |
| Fireview | Barrel stove with viewing window. | None stated | Large surface area. Cooker, heater and fireplace. | 180<br>360 | 118 lbs./20¼x18x16<br>279 lbs./26½x36½x22 | 4,000 cu. ft.<br>10,400 cu. ft. | 18"<br>36" | $245<br>$415 |
| Fisher | Steel, firebrick. Cast iron door. Airtight type. | 25 yrs. | Highly efficient with doors closed. | Grandpapa Bear<br>Grandma Bear | 475 lbs./33x30x30<br>425 lbs./33x25½x28 | 10,500 cu. ft.<br>9,000 cu. ft. | 24"<br>18" | $535<br>$515 |
| Frontier | Steel double door stepstove; airtight type. | Lifetime | Efficient. Every unit made by hand. | 74KR21101R<br>74KR21102R<br>74KR21103R | 330 lbs./27½x26x16½<br>380 lbs./29x28x18½<br>440 lbs./32½x30x20½ | 1,000 sq. ft.<br>1,400 sq. ft.<br>1,800 sq. ft. | 20"<br>22"<br>24" | $410<br>$445<br>$500 |
| Fyrtonden | Steel and firebrick combination heater. | 1 yr. | Superb design, sturdy and efficient. | A<br>B<br>C<br>D | 287 lbs./34x23 dia.<br>265 lbs./31x21 dia.<br>221 lbs./37x19 dia.<br>188 lbs./27x19 dia. | 7,000-9,000 cu. ft.<br>6,000-8,000 cu. ft.<br>5,000-7,000 cu. ft.<br>3,000-5,000 cu. ft. | 18"<br>16"<br>14"-16"<br>14" | $710<br>$690<br>$675<br>$570 |
| Garrison | Rolled steel and firebrick convertible stove. | None stated | Octagonal shape makes this a virtually indestructible, most attractive, sensible stove. | Garrison One<br>Garrison Two | 390 lbs./29½x32x21<br>245 lbs./25½x26x19 | 10,000 cu. ft.<br>7,500 cu. ft. | 24"<br>18" | $510<br>$405 |
| Gibraltar | Steel and firebrick. Racon window. | None stated | Airtight and efficient. | Gibraltar III | 350 lbs./32½x32x18½ | None stated | 24" | $455 |

| Combination Stoves | Construction | Guarantee | Mfr's Claims | The Models | Weight & Size (HxWxDepth) | Heating Capacity | Log Length | Price |
|---|---|---|---|---|---|---|---|---|
| Hydro-Temp | Steel plate and steel water jacket surrounding fire. Cast iron front and top. | 15 yrs. | Captures heat normally lost up chimney. | Hydro-Heater | 425 lbs./32x36x22 | 6,000 cu. ft. | 28" | $680 |
| Impression | Sheet steel, double-walled Franklin heater, airtight type. Optional blower. | 5 yrs. | Can be safely installed 10 inches from any wall surface. Stove jacket will not burn out. | Impression 5 | 300 lbs./31x30x31 | 1,500 sq. ft. without blower | 24" | $425 |
| Jotul | All cast iron, firebrick lined, airtight combination heater. | 2 yrs. | Fuel economy achieved by interior baffle plate. | No. 1<br>No. 4 | 183 lbs./33x19x19<br>286 lbs./41x22x22 | 7,000 cu. ft.<br>9,500 cu. ft. | 12"<br>14" | $520<br>$745 |
| King | Cast iron and steel. | None stated | Powerful radiant heater with doors closed; beams warmth and beauty with doors open. | 98-1800<br>98-1830 | 295 lbs./29½x32½x15<br>350 lbs./30½x38¾x16½ | None stated | 22"<br>28" | $495<br>$550 |
| Koco | Ball-shaped, wrought iron, free-standing stove. | 1 yr. | Conducts heat faster than cast iron, will not warp like box stoves. Spherical shape means easy ignition and efficient burning. | Koco | 175 lbs./45x26 dia. | None stated | 22" | $675 |
| Lange | All cast iron, firebrick lined, combination heater. | None stated | Airtight, powerful heater when closed. | 61 MF | 286 lbs./38x20½x19 | 5,000-7,000 cu. ft. | 16" | $585 |
| Logger | Cast iron and steel; grille attachment available. | 30 days | Convenience and radiant heat in an open fireplace. | OHF350 | 315 lbs./35x39x27 | 2-3 rooms | 20" | $415 |
| Morso | Cast iron and firebrick lined combination heater. | None stated | Regulated air flow when doors sealed. Biggest heat output in its class. | 1125 | 354 lbs./41x29x23 | 10,000 cu. ft. | 22" | $745 |
| Nashua | Boilerplate steel, firebrick lined. Radiant plus circulatory heat design. | Lifetime guarantee. | Alone in its field. Heats an entire house in five minutes flat. | Nashua Fireplace-1<br>Nashua Fireplace-2 | 385 lbs./29¾x27x33<br>525 lbs./32¾x34x33½ | 7,000-10,000 cu. ft.<br>12,000-16,000 cu. ft. | 18"<br>24" | $625<br>$720 |
| Old Timer | Welded boilerplate steel with cast iron doors. Firebrick lined, baffled. | 5 yrs. | Extremely airtight, rugged and efficient. Cooking area. | Old Timer Fireplace | 545 lbs./35¾x28½x30½ | An average home | 24" | $615 |
| Quaker | Steel and cast iron airtight type with glass doors. | Lifetime | Designed for warmth, safety and efficiency. | Moravian Parlor Stove | 510 lbs./35x31x21½ | 15,000 cu. ft. | 18" | $610 |
| Radke | All cast iron. | Dealer guarantee. | None stated | Imperial | approx. 250 lbs./37x38x25 | None stated | 20" | $170 |
| Sears Roebuck | Porcelained steel with glass doors. | Standard Sears home trial. | Our most efficient free-standing fireplace. | 42P8474N | 252 lbs./42½x41¼x29 | None stated | 20" | $460 |
| Scandia | All cast iron, firebrick lined; airtight when closed. | 5 yrs. | A rugged, economical, efficient Scandinavian stove. | Combi 200<br>Combi 300 | 286 lbs./40x23x23<br>300 lbs./38x22x29 | 3-5 rooms<br>4-6 rooms | 18"<br>20" | $310<br>$410 |
| Thermo-Control | Steel jacket with firebrick bottom. Airtight type. | 10 yrs. | Thermostatically controlled burner with door closed. Opens to enjoy fire. Durable and inexpensive. | Thermoplace<br>The Franklin | 240 lbs./31x24x24<br>190 lbs./28x30x24 | 800-1,200 sq. ft.<br>None stated | 20"<br>18" | $415<br>$345 |

## THE WOODSTOVE BUYER'S GUIDE [179]

| Combination Stoves | Construction | Guarantee | Mfr's Claims | The Models | Weight & Size (HxWxDepth) | Heating Capacity | Log Length | Price |
|---|---|---|---|---|---|---|---|---|
| Timberline | Airtight combination stove-fireplace; firebrick lined, baffled. Choice of legs or pedestal. | 5 yrs. | Rugged and efficient. | TSF<br>TLF | 480 lbs./33x26x26<br>568 lbs./33x29½x20 | 1,600 sq. ft.<br>2,400 sq. ft. | 24"<br>20" | $510<br>$560 |
| Trolla | Cast iron and firebrick; airtight type. | 1 yr. | Efficient heater with door closed. | 810 | 300 lbs./41x25x30 | 8,000 cu. ft. | 20" | $745 |
| Vermont Castings | All cast iron airtight type; fully baffled and thermostatically controlled. | 1 yr. | Highly efficient, beautiful cast iron heaters that convert to warm and friendly fireplaces. | Defiant<br>Vigilant | 340 lbs./34x36x22<br>245 lbs./32x32x25 | 8,000-10,000 cu. ft.<br>6,000- 8,000 cu. ft. | 24"<br>18" | $560<br>$460 |
| Warner | Boilerplate steel with cast iron door, baffled, airtight type. | Lifetime | Designed to be the most economical, efficient stove you can buy. | W 124 FP | 475 lbs./25x30x29 | 16,000 cu. ft. | 24" | $520 |
| Washington | Cast iron double door heaters with optional glass doors; airtight type. | 1 yr. | Economical and efficient stoves made the way they used to make them. | The Olympic<br>The 49'er<br>The Olympic Crest | 310 lbs./31x26x25<br>295 lbs./31¼x24¾x25<br>322 lbs./32½x36x24½ | None stated<br>None stated<br>None stated | 24"<br>20"<br>24" | $400<br>$425<br>$700 |
| Wonderwood | Steel and cast iron. | None stated | Baffle creates longer flame path, greater efficiency. | The Franklin | 195 lbs./27x38x21 | None stated | 28" | $250 |
| Zodiac | Cast iron, sheet metal and chrome, contemporary-styled round heater. | 1 yr. | A unique design you will never cease to admire. | The Zodiac | 250 lbs./38x26 dia. | None stated | 12" | $515 |

[180] HEATING WITH WOOD

## SOME FINAL NOTES

First, a word about heating capacity.

These figures, provided by the woodstove manufacturers, are the most general estimates of a woodburner's capabilities. The ability of a stove to heat a given area depends more on the configuration and contents of the cubic feet of space involved than on the stove itself.

For your convenience, we have also included distilled statements of stove manufacturers' claims for their products. These should be viewed as manufacturers' claims and nothing more, to be verified by personal inquiry. With the bewildering variety of woodstoves on the market, the hard sell is something staged. Take your time examining the woodheaters that interest you in the dealer's showroom so that you can make an informed decision about the type of woodburner that's best for you.

Every effort has been made to assure that this guide is accurate and complete. However, neither the publisher nor the author can be responsible for omissions, typographical or factual errors, manufacturers' changes in specifications provided for the compilation of this guide, or variations from the suggested retail prices listed here, which will certainly occur not only from region to region of the country, but also from stove shop to stove shop within your own community.

For more information about any of the woodburners listed here, contact the manufacturers and regional distributors listed below:

**AFS,** Intercontinental Building Products, 620 East Main St., Orange, Mass. 01364.

**All Nighter** Stove Works Inc., 80 Commerce St., Glastonbury, Conn. 06033.

**Alpiner,** Lyons Supply Co., 1 Perimeter Rd., Manchester, N.H. 03108.

**Arctic,** Washington Stove Works, Box 687, Everett, Wash. 98206.

**Ardenne,** Lyons Supply Co., 1 Perimeter Rd., Manchester, N.H. 03108.

**Ashley,** P.O. Box 128, Florence, Ala. 35630.

**Atlanta** Stove Works Inc., P.O. Box 5254, Atlanta, Ga. 30307.

**Atlantic,** Portland Stove Foundry Co., 57 Kennebec St., Portland, Maine 04104.

**Autocrat,** New Athens, Ill. 62264.

**Bellway,** Perley C. Bell, Grafton, Vt. 05146.

**Birmingham** Stove and Range Co., P.O. Box 2647, Birmingham, Ala. 35202.

**Bullard** Manufacturing Co., 82 Learney Ave., Liverpool, N.Y. 13088.

**Canadian Stepstove,** New Hampshire Stove Co., 19 North Main St., Wolfeboro, N. H. 03894.

The **Cawley/LeMay** Stove Co., P.O. Box 561, Boyertown, Pa. 19512.

**Chappee,** Preston Distributing Co., 2 Whidden St., Lowell, Mass. 01852.

**Charmaster** Products Inc., 2307 Highway 2 West, Grand Rapids, Minn. 55744.

**Cherokee,** Gregg Distributing Co., Box 37, Stokesdale, N.C. 27357.

**Combo** Furnace Co., 1707 West 4th St., Grand Rapids, Minn. 55744.

**Comforter,** Abundant Life Farm, Box 175, Lochmere, N.H. 03252.

**Controlled Combustion Systems,** 1978 Washington St., Hanover, Mass. 02339.

**Country Craftsmen,** Box 3333, Stanta Rosa, Calif. 95402.

**Culvert Queen,** L.W. Gay Stoveworks, 156 Vernon Rd., Brattleboro, Vt. 05301.

Sam **Daniels** Co. Inc., Box 868, Montpelier, Vt. 05602.

**De Dietrich,** The Burning Log, P.O. Box 438, Lebanon, N.H. 03766

**Dover** Stove Co., Box 217, Sangerville, Maine 04479.

**Dynamite** Stove Co., RD 3, Montpelier, Vt. 05602.

**Efel Kamina,** Southport Stoves Inc., 959 Main St., Stratford, Conn. 06497.

**El Fuego** Co., 26 Main St., Oakville, Conn. 06779.

**Elm,** Vermont Iron Stove Works Inc., Warren, Vt. 05674.

**Energy Harvesters** Corp., Box 19, Fitzwilliam, N.H. 03447.

**Fatsco** Stoves, 251 Fair Ave., Benton Harbor, Mich. 49022.

**Findlay Oval,** Elmira Stove Works, 22 Church St., West Elmira, Ontario, Canada.

**Firemagic,** Del Gilbert Co., RFD 2, Highway 107, Laconia, N.H. 03246.

**Fireview,** Tate Equipment Inc., Horseheads, N.Y. 14845.

**Fisher** Stoves, 504 South Main St., Concord, N.H. 03301.

**Fisher's,** Rte. 1, Box 63A, Conifer, Colo. 80433.

**Fjord,** Lyons Supply Co., 1 Perimeter Rd., Manchester, N.H. 03108.

**Free Flow** Stove Works, South Stratford, Vt. 05070.

**Free Heat Machine,** Brookfield Fireside, Rte. 7, Brookfield, Conn. 06804.

**Frontier,** Montgomery Ward.

**Fyrtonden,** Bow and Arrow Stove Co., 14 Arrow St., Cambridge, Mass. 02138.

**Garrison** Stove Works, Box 412, Claremont, N.H. 03743.

**Gibraltar,** Sierra Stoves, 1 Appletree Square, Minneapolis, Minn. 55420.

**Hearth Mate,** C&D Distributors Inc., Box 766, Old Saybrook, Conn. 06475.

**Heatilator** Co., Mt. Pleasant, Iowa 52641.

**Heatscreen Plus,** Sears, Roebuck.

**Hede,** Lyons Supply Co., 1 Perimeter Rd., Manchester, N.H. 03108.

**Hinckley Shaker,** Hinckley Foundry & Marine, 13 Water St., Newmarket, N.H. 03857.

**Home Warmers,** New Hampshire Woodstoves Inc., P.O. Box 310, Plymouth, N.H. 03264.

**Hoval Boiler,** Arotek Corp., 1703 East Main St., Torrington, Conn. 06790.

**Hunter,** Integrated Thermal Systems, 379 State St., Portsmouth, N.H. 03801.

**Huntsman,** Sears, Roebuck.

**Hydroplace,** Ridgeway Steel Fabricators Inc., Box 382, Ridgeway, Pa. 15853.

**Hydro-Temp,** RD 1, Box 257, Narvon, Pa. 17555.

**Impression,** KNT, Box 25, Hayesville, Ohio 44838.

**Independence,** L.W. Gay Stoveworks, 156 Vernon Rd., Brattleboro, Vt. 05301.

**Jotul,** The Burning Log, Box 438, Lebanon, N.H. 03766.

**Kachelofen,** Ceramic Radiant Heat, Lochmere, N.H. 03252.

**Kerr Controls Ltd.,** 9 Circustime Rd., South Portland, Maine 04106.

**Kickapoo** Stove Works, Box 127, La Farge, Wis. 54639.

**King,** Martin Industries, P.O. Box 128, Florence, Ala. 35630.

**Koco,** Scandia Wood Stoves Inc., 174 Old York Rd., New Hope, Pa. 18938.

**Koppe,** Finest Stove Imports Inc., P.O. Box 1733, Silver Spring, Md. 20902.

**Lakewood,** Woodman Associates, Box 626, Wakefield, N.H. 03872.

**Lange,** Svendborg Co. Inc., Box 5, Hanover, N.H. 03755.

**Leyden Hearth,** Leyden Energy Conservation Corp., Brattleboro Rd., Leyden, Mass. 01337.

**Locke Stove** Co., 114 West 11th St., Kansas City, Mo. 64105.

**Logger** Stove Corp., 1104 Wilso Drive, Baltimore, Md. 21223.

**Logwood,** Marathon Heater Co., Box 265, RD 2, Marathon, N.Y. 13803.

**Longwood,** Masi Plumbing and Heating, 36 Otterson St., Nashua, N.H. 03060.

**Lynndale,** Integrated Thermal Systems, 379 State St., Portsmouth, N.H. 03801.

**Martin** Industries, Box 128, Florence, Ala. 35630.

**Majestic,** Lyons Supply Co., 1 Perimeter Rd., Manchester, N.H. 03108.

**Monarch** Kitchen Appliances, 316 South Perry St., Johnstown, N.Y. 12095.

**Monarch,** Lyons Supply Co., 1 Perimter Rd., Manchester, N.H. 03108.

**Montgomery Ward,** Montgomery Ward stores.

**Morso,** Inglewood Stove Co., Rte. 4, Woodstock, Vt. 05091.

**Nashua,** Heathdelle Sales Associates Inc., Rte. 3, Meredith, N.H. 03253.

**Newmac,** Arotek Corp., 1703 East Main St., Torrington, Conn. 06790.

**Norflame,** Lyons Supply Co., 1 Perimeter Rd., Manchester, N.H. 03108.

**Norman,** Lyons Supply Co., 1 Perimeter Rd., Manchester, N.H. 03108.

**Northeaster,** Solar Wood Energy Corp., Fall Rd., East Lebanon, Maine 04027.

**Norwester,** Washington Stove Works, Box 687, Everett, Wash. 98206.

**Old Timer,** Midwest Stoves Inc., P.O. Box 1704, Mt. Vernon, Ill. 62864.

**Olympic,** Washington Stove Works, Box 687, Everett, Wash. 98206.

**Oneida Royal,** Edward R. Stephen Co. Inc., 78 Franklin St., Somerville, Mass. 02145.

**Passat** USA, Box 37, East Kingston, N.H. 03848.

**Pillsbury** Stove Works, 84 Hathaway St., Providence, R.I. 02907.

**Pioneer** Lamps and Stoves, 71 Yesler Way, Seattle, Wash. 98104.

**Prospector Gold-Mark Industries Inc.,** P.O. Box 350 East Broadway, Monticello, N.Y. 12701.

**Quaker,** Woodman Associates, Box 626, Wakefield, N.H. 03872.

**Radke** Imports, P.O. Box 128, Emmett, Idaho 83617.

**Ram** Forge, Brooks, Maine 04921.

**Reginald,** S/A Imports, 730 Midtown Plaza, Syracuse, N.Y. 13210.

**Riteway** Manufacturing Co., P.O. Box 6, Harrisonburg, Va. 22801.

**Scandia,** Preston Fuel, Whidden St., Lowell, Mass. 01850.

**Sears, Roebuck,** Sears, Roebuck stores.

**Sevca** Stoveworks, Box 477, Saxtons River, Vt. 05154.

**Shenandoah,** P.O. Box 839, Harrisonburg, Va. 22801.

**Sierra** Stoves, 1 Appletree Square, Minneapolis, Minn. 55420.

**Spaulding,** Novatek Inc., 79R Terrace Hill Ave., Burlington, Mass. 01803.

**Stanley,** Inglewood Stove Co., Rte. 4, Woodstock, Vt. 05091.

**Styria,** The Merry Music Box, 20 McKown St., Boothbay Harbor, Maine 04538.

**Suburban** Manufacturing Co., P.O. Box 399, Dayton, Tenn. 37321.

**Sunshine** Stove Works Inc., Norridgewock, Maine 04957.

**Tarm,** Tekton, Conway, Mass. 01341.

**Tasso,** Tekton, Conway, Mass. 01341.

**Tempwood,** Mohawk Industries, 173 Howland Ave., Adams, Mass. 01220.

**Thermo-Control** Woodstoves, Box 640, Cobleskill, N.Y. 12043.

**Thermo Pride,** Thermo Products Inc., P.O. Box 217, North Judson, Ind. 46366.

**Thriftchanger,** Sturges Heat Recovery Co., Box 397, Stone Ridge, N.Y. 12484.

**Tiba,** Svendborg Co. Inc., Box 5, Hanover, N.H. 03755.

**Timberline,** Energysavers Inc., Parade Rd. at Rte. 3, Meredith, N.H. 03253.

**Tirolia,** 169 Dunning Rd., Middletown, N.Y. 10940.

**Trolla,** Lyons Supply Co., 1 Perimeter Rd., Manchester, N.H. 03108.

**Ulefos,** Scandia Wood Stoves Inc., 174 Old York Rd., New Hope, Pa. 18938.

**Valley Comfort,** Woodburning Specialties, P.O. Box 5, North Marshfield, Mass. 02059.

**Vermont** Castings Inc., Box 126, Randolph, Vt. 05060.

**Vermont Downdrafter,** Vermont Woodstove Co., P.O. Box 1016, Bennington, Vt. 05201.

**Volcano,** Anchor Industries, Rte. 12, Box 63, Fitzwilliam, N.H. 03447.

**Waverly** Heating Supply Co., 117 Elliot St., Beverly, Mass. 01915.

**Wamsler,** Logger Stove Corp., 1104 Wilso Drive, Baltimore, Md. 21223.

**Warner** Woodstove Co., Box 292, Warner, N.H. 03278.

**Washington,** Vermont Distributors, 11 Maple St., Essex Junction, Vt. 05452.

**Weso,** Ceramic Radiant Heat, Lochmere, N.H. 03252.

**Wonderwood,** U.S. Stove Co., South Pittsburg, Tenn. 37380.

**Wood King,** Martin Industries, Box 128, Florence, Ala. 35630.

**Woodsman,** Crendall-Hicks Co., Rte. 9, Southborough, Mass. 01772.

**Yankee** Woodstoves, P.O. Box 7, Bennington, N.H. 03442.

**Yukon,** Moutton Climate Control, Portsmouth, N.H. 03801.

**Zodiac,** Washington Stove Works, Box 687, Everett, Wash. 98206.

# (2) The Chain Saw Buyer's Guide

KEY to Standard Features

BAR TYPES
S-nose • Sprocket nose
H-nose • Hard nose
R-nose • Roller nose

ENGINE MECHANISMS
EI • Electronic Ignition
Dec • Decompressor for starting

OILING METHODS
AMO • Automatic and Manual Oiling
AO • Automatic Oiling
MO • Manual Oiling

SAFETY FEATURES
CB • Chain Brake
KN • Knuckle Guard
CGC • Chain Guard Catcher
RHG • Right Hand Guard

MISCELLANEOUS FEATURES
AV • Anti-Vibration
EHH • Electronically Heated Handles
Sqk • Spike for steadying

HERE IS A REPRESENTATIVE list of major chain saw models and options available in most parts of the United States today. For each manufacturer, three saws have been listed: one representing a backyard mini-saw model, another a lightweight, intermediate-size production model, and the third a heavy-duty professional model. In almost all cases, the manufacturers also offer other chain saws that fall within these classes, and these models may feature different combinations of options. The names and locations of local dealers can be obtained by checking your Yellow Pages.

| Chain Saw | Distributor | Guarantee | Models | Engine Displacement | Weight (without bar and chain) | Bar Length | Standard Features | Price (suggested retail price) |
|---|---|---|---|---|---|---|---|---|
| Allis-Chalmers | Allis-Chamlers 7176 Morgan Rd. Liverpool, N.Y. 13088 | 90 days | 65 75 no heavy-duty model | 1.9 cu. in. (30cc) 2.3 cu. in. (38cc) | 8 lbs. 8.9 lbs. | 12" 14" | MO MO | $145 $180 |
| Echo-Kioritz | Timberland Machines 10 No. Main St. Lancaster, N.H. 03584 | 1 yr. for homeowner 90 days for commercial use | 315 452 VL 702EVL | 1.8 cu. in. (30cc) 2.7 cu. in. (44cc) 4.3 cu. in. (71cc) | 9.2 lbs. 13.2 lbs. 18 lbs. | 12" 16" 20" | AO, AR, S-nose AMO, AV, S-nose AMO, Dec, El, S-nose | $145 $265 $385 |
| Homelite | Homelite Homelite Distribution 60 Chapin Rd. Pinebrook, N.J. 07058 | 1 yr. for homeowner 90 days for commercial use | XL XL12 Super XL925 | 1.6 cu. in. (26cc) 3.3 cu. in. (54cc) 5.0 cu. in. (82cc) | 7.1 lbs. 13.3 lbs. 16.7 lbs. | 10" 16"-24" 15"-36" | AO R-nose, Spk MO, R-nose AMO, El, R-nose, Spk | $115 $210-$270 $420-$455 |

[186]

## THE CHAINSAW BUYER'S GUIDE [187]

| Chain Saw | Distributor | Guarantee | Models | Engine Displacement | Weight (without bar and chain) | Bar Length | Standard Features | Price (suggested retail price) |
|---|---|---|---|---|---|---|---|---|
| Husqvarna | Jesse F. White<br>Route 16<br>Mendon, Mass. 01756 | 90 days | 32VR<br>162SG<br>285 | 2.0 cu. in. (32cc)<br>3.5 cu. in. (62cc)<br>5.2 cu. in. (85cc) | 14 lbs.<br>16 lbs.<br>21 lbs. | 12"-14"<br>16"-20"<br>20"-32" | AO, AV, CGC, KN, R-nose<br>AO, AV, CGC, EL, EHH, S-nose, Spk<br>AO, AV, S-nose, Spk | $205-$220<br>$465-$480<br>$515-$545 |
| Jonsereds | Tilton Equipment<br>Route 1<br>Rye, N.H. 03870 | 60 days | 361<br>49SP<br>70E<br>(replaces 66E) | 2.1 cu. in. (35cc)<br>3.0 cu. in. (49cc)<br>4.2 cu. in. (69cc) | 6.8 lbs.<br>12.5 lbs.<br>18 lbs. | 14"<br>15"-18"<br>16"-24" | AMO, CGC, R-nose<br>AO AV, CGC, S-nose<br>AO, AV, CGC, El, S-nose | $150<br>$275-$290<br>$395-$410 |
| John Deere | John Deere Co.<br>P.O. Box 4949<br>Syracuse, N.Y. 13221 | 1 yr. | 30<br>50V<br>70V | 1.8 cu. in. (30cc)<br>2.7 cu. in. (44cc)<br>4.3 cu. in. (71cc) | 9.9 lbs.<br>13.3 lbs.<br>18 lbs. | 12"-14"<br>16"<br>16"-24" | AO, S-nose<br>AO, AV, CB, KN, Spk, S-nose<br>AO, AV, CB, KN, Spk, S-nose | $185-$200<br>$250<br>$360-$375 |
| Lombard | R. D. Faulkner<br>146 Parkway South<br>Brewer, Me. 04412 | 2 yrs. for homeowner<br>90 days for commercial use | CS2600<br>Comango<br>Super Lightning | 2.2 cu. in. (35cc)<br>4.2 cu. in. (70cc)<br>4.2 cu. in. (70cc) | 7.7 lbs.<br>13 lbs.<br>15 lbs. | 10"-16"<br>16"-24"<br>16"-24" | AO, S-nose<br>AMO, S-nose<br>AMO, AV, S-nose | $150-$175<br>$275-$300<br>$275-$310 |
| McCulloch | McCulloch of New England<br>Reading, Mass. 01867 | 1 yr. for homeowner<br>60 days for commercial use | PM310<br>PM610<br>PM700 | 2.1 cu. in. (35cc)<br>3.7 cu. in. (60cc)<br>4.3 cu. in. (70cc) | 10 lbs.<br>15 lbs.<br>15.3 lbs. | 14"<br>16"-28"<br>16"-28" | AMO, AV, CB, CGC, El, KN, RHG, S-nose<br>AMO, AV, CB, CGC, El, KN, S-nose<br>AMO, CB, CGC, Dec, El, KN, S-nose | $175<br>$250-$275<br>$350-$375 |
| Montgomery Ward | only by catalogue | 90 days | 24050<br>24056<br>30020A | 1.9 cu. in. (31cc)<br>2.3 cu. in. (44cc)<br>3.6 cu in. (57cc) | 8 lbs.<br>10 lbs.<br>15 lbs. | 10"<br>16"<br>24" | AO, AV, CB<br>AO, AV, CB, S-nose<br>AMO, R-nose | $85<br>$210<br>$375 |
| Olympic | Tilton Equipment<br>Route 1<br>Rye, N.H. 03870 | 60 days | 240<br>251<br>Super 480 | 2.3 cu. in. (38cc)<br>3.0 cu. in. (49cc)<br>5.0 cu. in. (81cc) | 9 lbs.<br>13.6 lbs.<br>17.8 lbs. | 14"<br>16"-20"<br>16"-24" | AO, KN, RHG, S-nose<br>AO, KN, RHG<br>AO, KN, RHG, S-nose | $175<br>$225-$250<br>$350-375 |
| Partner | R. D. Faulkner<br>146 Parkway South<br>Brewer, Me. 04412 | 90 days | R517T<br>P70<br>P100 | 3.55 cu. in. (55cc)<br>4.5 cu. in. (70cc)<br>6.1 cu. in. (100cc) | 16 lbs.<br>14 lbs.<br>17 lbs. | 15"-22"<br>16"-26"<br>18"-26" | AO, AV, El, S-nose, Spk<br>AO, AV, Dec, El, S-nose, Spk<br>AO, AV, El, S-nose, Spk | $390-$400<br>$460-$475<br>$575-$595 |
| Pioneer | Crandall Hicks Co.<br>Route 9<br>Southboro, Mass. 01772 | 90 days | 1074<br>1200A<br>P50 | 3.1 cu. in. (51cc)<br>3.5 cu. in. (57cc)<br>5.0 cu. in. (82cc) | 10.8 lbs.<br>13.8 lbs.<br>16.5 lbs. | 14"<br>16"<br>16" | AO, Dec, S-nose<br>AMO, Dec, S-nose<br>AMO, Dec, S-nose | $200<br>$300<br>$385 |
| Poulan | R. E. Jarvis<br>Route 9<br>Fayville, Mass. 01745 | 90 days | Micro 25<br>S25DA<br>42000CVA | 2.0 cu. in. (33cc)<br>2.3 cu. in. (38cc)<br>4.2 cu. in. (69cc) | 7.5 lbs.<br>8.6 lbs.<br>16 lbs. | 10"-12"<br>14"<br>16"-30" | AO<br>AMO, H-nose<br>AO, AV, El | $100-$135<br>$195<br>$410-$445 |
| Sears Craftsman | only by catalogue | 90 days | 35201<br>35851<br>35871 | 2.0 cu. in. (33cc)<br>3.5 cu. in. (57cc)<br>4.5 cu. in. (74cc) | 9 lbs.<br>11 lbs.<br>14 lbs. | 10"<br>20"<br>23" | AO, KN<br>AO, AV, S-nose, Spk<br>AMO, AV, S-nose, Spk | $100<br>$380<br>$450 |
| Skil | Skil Corp.<br>Box 612<br>Hampton, N.H. 03842 | 1 yr. | 1614<br>1632<br>1646 | 2.2 cu. in. (36cc)<br>3.4 cu. in. (56cc)<br>4.0 cu. in. (66cc) | 9 lbs.<br>12.3 lbs.<br>14.1 lbs. | 14"<br>16"<br>20" | AMO, AV, CB, El, Spk<br>AO, CB, S-nose, Spk<br>AO, AV, CB, S-nose | $185<br>$295<br>$385 |
| Solo Motor Inc. | Doe Ag-Sales<br>Wescott Rd.<br>Harvard, Mass. 01451 | 90 days | 600<br>650<br>642 | 2.0 cu. in. (32cc)<br>3.8 cu. in. (62cc)<br>6.5 cu. in. (106cc) | 11 lbs.<br>15 lbs.<br>16.5 lbs. | 12"<br>16"-20"<br>17"-32" | AO, CB, KN, RHG, S-nose, Spk<br>AO, AV, KN, H-nose, Spk<br>AMO, AV, H-nose, Spk | $215<br>$400<br>$425-$475 |
| Stihl | Hampton Equipment<br>Lancaster, N.H. 03584 | 90 days | 020AV<br>041AV<br>051AVE | 2.0 cu. in. (32cc)<br>3.7 cu. in. (51cc)<br>5.3 cu. in. (87cc) | 9.7 lbs.<br>18 lbs.<br>25 lbs. | 12"-16"<br>16"-24"<br>17"-36" | AO, AV, S-nose<br>AO, AV, S-nose<br>AO, AV, El, S-nose | $275-$295<br>$400-$415<br>$510-$550 |
| Tilton Equipment | Tilton Equipment<br>Route 1<br>Rye, N.H. 03870 | 60 days | Lil'T<br>Tilton 420<br>no heavy-duty model | 2.2 cu. in. (36cc)<br>4.2 cu. in. (69cc) | 7.8 lbs.<br>13.8 lbs. | 12"-14"<br>16"-24" | AO, S-nose<br>AMO | $150-$165<br>$250-$275 |

# (3) Tree Identification and Heat Value Guide

Numbers after tree names indicate heating value in million BTUs per cord of air-dry wood.

---

**Best Firewood** | **Good Firewood** | **Poor Firewood**

### Locust, Black  26.5

Leaves are a compound with six to 20 leaflets. Bark rough, deeply furrowed. Branches armed with stout spines one-half to one inch long. Flowers are large, white, showy, in clusters, and develop in June. Fruit is a flat pod about three inches long, containing several hard, dark brown seeds. Tree up to 60 feet high.

### Maple, Red  19.1

Leaves are five-lobed or less, with sharp angles between the lobes and sharp irregular teeth, pale beneath. Bark is grey, smooth on young trees, separating and forming long scales on older trees. Flowers are red or yellow and appear in April before the leaves. Tree up to 100 feet high and three feet in diameter.

### Hemlock  15.0

Needles are short, about one-half inch long with two silvery lines beneath. Resembles balsam, but needles more irregularly arranged and shorter, not fragrant. Bark is thick, reddish, developing deep, coarse ridges on older trees. Cones are round and small. Tree up to 80 feet high and three feet in diameter.

### Hickory, Shagbark  25.4

Usually five leaflets, with three upper leaflets much larger, toothed, with a minute tuft of hairs on each tooth. Bark on older trunks flaking into long plates. Nuts with thick husks readily splitting away when ripe, variable in size and shape on different trees.

### Tamarack (Pine)  19.1

Needles are short in circular clusters of 10 to 30, one-half to one inch long. In autumn they turn yellow and fall, giving the tree a dead appearance. Bark is reddish-brown and scaly. This is the only member of the pine family to lose its leaves in winter.

### Spruce, Red  15.0

Needles are one-half to three-quarters of an inch long, bright green, four-sided, sharp-pointed. Twigs are covered with few to many very small hairs. Bark is thick, scaly, reddish-brown. Cones are nearly round or egg-shaped, one to two inches long. Large, slender tree up to 100 feet high and three feet in diameter.

## Best Firewood

### Hophornbeam (Birch)   24.7

Leaves are sharply toothed, one to four inches long, downy underneath. Bark is thin, grey, with long flaky scales which lift up from the edges and cling in the middle. Fruit is a hoplike cluster of bladdery sacs, each enclosing a small, flat nut. Tree usually not more than 30 feet high.

### Oak, White   23.9

Leaves are smooth underneath, with prominent lobes, four to 10 in number, rounded, variable in depth of cutting, not markedly broader above middle than below. Bark is grey, thin and flaky. Cup of acorn much shorter than nut, covered with wart-like scales.

### Beech   21.8

Leaves are coarsely toothed with prominent parallel veins terminating in the teeth, three to five inches long. Dry leaves remain on the smaller trees through the winter. Bark is hard, smooth, steel grey.

## Good Firewood

### Cherry, Black   18.5

Leaves are up to four inches long with teeth that are somewhat rounded. Bark is dark red-brown, smooth on young trees, later breaking into scales. Flowers are white, in elongated clusters up to four inches long, blooming in late May or June. Fruit is black. Tree up to 90 feet high, but usually not very large.

### Pine, Pitch   18.5

Needles are in clusters of threes, one-and-one-half to four inches long, stiff, yellowish-green. Bark is thick and rough, deeply furrowed, dark reddish-brown. Clusters of needles often grow out of the bark. Cones are short, one to three inches long, often clustered. Tree up to 50 feet high, scrubby, quick tapering.

### Birch, White   18.2

Leaves are broad-toothed, rounded at base, one to four inches long. Bark is chalky white, outer layers separate into thin strips. Twigs are dark and roughened with white spots. Flowers are brownish and develop in May. Tree up to 70 feet high and three feet in diameter.

## Poor Firewood

### Butternut (White Walnut)   14.3

Leaves are a compound with seven to 17 leaflets, large, often one-and-one-half feet long, hairy underneath. Leaf stalks and young twigs are covered with sticky hairs. Bark is light grey, smooth when young, later separating into flat, braided ridges. Tree occasionally up to 90 feet high.

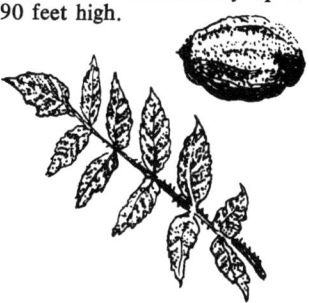

### Cherry, Pin   14.2

Leaves are narrow, taper-pointed, finely and evenly toothed, the teeth tipped with knobbed hairs. Bark is thin, reddish-brown. Flowers are white with round petals, in clusters, all arising from the same point on the stem. Small tree up to 30 feet high and one foot in diameter.

### Aspen, Trembling   14.1

Leaves are rounded with a short, abrupt point, finely and evenly toothed. Leaf stalks are slender and flattened, which causes leaves to tremble easily in the wind. Bark on young trees is smooth, green with dark patches, darker and rough on older trees. Tree up to 60 feet high.

## Best Firewood

### Maple, Sugar 21.8

Leaves are usually five-lobed, rounded between the lobes and without sharp teeth, pale underneath. Bark on young trees is smooth, on old trees grey and deeply furrowed. Flowers are yellow-green without petals. Fruit opposite twigs, matures in September. Large tree up to 100 feet high and three feet in diameter.

### Oak, Red 21.7

Leaves are smooth on both surfaces, with mainly three pairs of lateral lobes. Their sides tend to diverge and form a V-shaped space between lobes. Bark is greyish-brown, hard and with deep furrows on older trees. Acorn cup is shallow, much shorter than the acorn. Large tree up to 70 feet high and three feet in diameter.

## Good Firewood

### Maple, Silver 17.9

Leaves are deeply cut between the lobes, silvery beneath. Twigs are usually long and drooping. Bark is smooth on young trees, grey; on old trees it becomes furrowed and scaly. Flowers are red or yellow with no petals, appearing in late March or April. Fruit matures in early summer. Tree up to 100 feet high and three feet in diameter.

### Pine, Norway 17.8

Needles are long, in clusters of twos, three to six inches long, sharp-pointed. Bark is reddish-brown, thick, divided by shallow furrows into broad, flat ridges. Twigs are stout and roughened. Cones are two to three inches long, seated directly on the twig. Large tree up to 80 feet high and three feet in diameter.

## Poor Firewood

### Fir, Balsam 13.5

Needles are one-half to one-and-one-half inches long, pale green with two silvery white lines beneath, soft and fragrant when crushed. Bark is smooth even on old trees, often with pitch-filled blisters. Tree, when well developed, with a spire-like shape up to 60 feet high and two feet in diameter.

### Willow, Black 13.5

Leaves are long and narrow, deep green on both sides, often tapering to a curved point, three to six inches long, finely toothed. Bark of larger trees is rough and scaly. Small tree 10 to 40 feet high.

# TREE IDENTIFICATION GUIDE [191]

## Best Firewood

### Birch, Yellow  21.3

Leaves are heart-shaped at base, sharply double-toothed, two to three inches long. Bark is yellowish or silvery grey, detaching in thin layers. Twigs are smooth with wintergreen flavor. Flowers are brownish, fruit matures in July. Large tree up to 60 feet high and three feet in diameter.

## Good Firewood

### Elm, White  17.7

Leaves are lopsided at base, with prominent parallel veins, toothed, two to four inches long, slightly rough on upper surface, smooth on lower surface. Bark is thick, greyish-brown, becoming rough and deeply furrowed on older trees. Twigs are smooth and slender. Large, fast-growing tree up to 100 feet high and six feet in diameter.

## Poor Firewood

### Pine, White  13.3

Needles are in clusters of fives, three to five inches long, very slender, falling at the end of the second season. Bark on young trees is smooth, on older trees dark and rough. Cones form in June. Large tree frequently up to 100 feet high and three feet in diameter.

### Ash, White  20.0

Leaves are smooth or with occasional scattered hairs, a compound with mainly five to seven leaflets up to four inches long, smooth underneath, often having coarse teeth, each leaflet borne on a distinct but short stalk. Bark on young trees is smooth, on older trees dark brown. Tall straight-trunked tree up to 80 feet high.

### Birch, Grey  17.5

Leaves are rather small, triangular in outline with long tapering tips, finely double-toothed with a distinctive unpleasant taste. Bark is close, not peeling, chalky white with dark markings. Twigs are dark, roughened with whitish dots. Flowers are brownish and develop in May. Tree not more than 30 feet high or one foot in diameter.

### Basswood  12.6

Leaves are sharply toothed, with slender-tipped teeth, unequally heart-shaped at base, often large, up to eight inches long. Bark is smooth and grey, becoming thick and ridged on older trees. Flowers are about one-half inch across with five yellowish petals. Large tree up to 100 feet high and three feet in diameter.

# The Last Word

## ODE FOR A WOODBURNER

BEECHWOOD fires are bright and clear,
  If the logs are kept a year.

CHESTNUT'S only good, they say,
  If for long it's laid away.

BIRCH and FIR logs burn too fast,
  Blaze up bright and do not last.

ELM wood burns like a churchyard mould;
  Even the very flames are cold.

POPLAR gives a bitter smoke,
  Fills your eyes and make you choke.

APPLE wood will scent your room
  With an incense like perfume.

OAK and MAPE, if dry and old,
  Keeps away the winter's cold.

But ASH wood wet, and ash wood dry,
  A king shall warm his slippers by.

*—Anonymous*

TODAY, IN THE CLOSING days of the twentieth century, our nation is blessed with a bounty of rich green forests that hold the promise of meeting our energy needs for countless years to come.

Yet without proper management and care, there is no guarantee that these forests will be here for the benefit of our children, and our children's children. Forests once sprawled over now-desert areas of Africa and the Middle East. Cut down for fuel and never replenished, they died.

Use the forest well, but use it wisely. For every tree you cut down, see to it there is another one planted, so that future generations may enjoy the forests as you do. The forests are our hope for the future.